T0325143

An Analysis of

Rachel Carson's

Silent Spring

Nikki Springer

Published by Macat International Ltd
24:13 Coda Centre, 189 Munster Road, London SW6 6AW.

Distributed exclusively by Routledge
2 Park Square, Milton Park, Abingdon, Oxon OX14 4RN
711 Third Avenue, New York, NY 10017, USA

Routledge is an imprint of the Taylor & Francis Group, an informa business

www.macat.com
info@macat.com

Cataloguing in Publication Data
A catalogue record for this book is available from the British Library.
Library of Congress Cataloguing-in-Publication Data is available upon request.
Cover illustration: Kim Thompson

ISBN 978-1-912302-35-2 (hardback)
ISBN 978-1-912127-45-0 (paperback)
ISBN 978-1-912281-23-7 (e-book)

Notice
The information in this book is designed to orientate readers of the work under analysis,
to elucidate and contextualise its key ideas and themes, and to aid in the development
of critical thinking skills. It is not meant to be used, nor should it be used, as a
substitute for original thinking or in place of original writing or research. References and
notes are provided for informational purposes and their presence does not constitute
endorsement of the information or opinions therein. This book is presented solely for
educational purposes. It is sold on the understanding that the publisher is not engaged
to provide any scholarly advice. The publisher has made every effort to ensure that
this book is accurate and up-to-date, but makes no warranties or representations with
regard to the completeness or reliability of the information it contains. The information
and the opinions provided herein are not guaranteed or warranted to produce particular
results and may not be suitable for students of every ability. The publisher shall not be
liable for any loss, damage or disruption arising from any errors or omissions, or from
the use of this book, including, but not limited to, special, incidental, consequential or
other damages caused, or alleged to have been caused, directly or indirectly, by the
information contained within.

CONTENTS

THE MACAT LIBRARY

The Macat Library is a series of unique academic explorations of seminal works in the humanities and social sciences – books and papers that have had a significant and widely recognised impact on their disciplines. It has been created to serve as much more than just a summary of what lies between the covers of a great book. It illuminates and explores the influences on, ideas of, and impact of that book. Our goal is to offer a learning resource that encourages critical thinking and fosters a better, deeper understanding of important ideas.

Each publication is divided into three Sections: Influences, Ideas, and Impact. Each Section has four Modules. These explore every important facet of the work, and the responses to it.

This Section-Module structure makes a Macat Library book easy to use, but it has another important feature. Because each Macat book is written to the same format, it is possible (and encouraged!) to cross-reference multiple Macat books along the same lines of inquiry or research. This allows the reader to open up interesting interdisciplinary pathways.

To further aid your reading, lists of glossary terms and people mentioned are included at the end of this book (these are indicated by an asterisk [*] throughout) – as well as a list of works cited.

Macat has worked with the University of Cambridge to identify the elements of critical thinking and understand the ways in which six different skills combine to enable effective thinking.
Three allow us to fully understand a problem; three more give us the tools to solve it. Together, these six skills make up the **PACIER** model of critical thinking. They are:

ANALYSIS – understanding how an argument is built
EVALUATION – exploring the strengths and weaknesses of an argument
INTERPRETATION – understanding issues of meaning

CREATIVE THINKING – coming up with new ideas and fresh connections
PROBLEM-SOLVING – producing strong solutions
REASONING – creating strong arguments

To find out more, visit **WWW.MACAT.COM.**

CRITICAL THINKING AND *SILENT SPRING*

Primary critical thinking skill: REASONING
Secondary critical thinking skill: CREATIVE THINKING

Rachel Carson's 1962 *Silent Spring* is one of the few books that can claim to be epoch-making. Its closely reasoned attack on the use of pesticides in American agriculture helped thrust environmental consciousness to the fore of modern politics and policy, creating the regulatory landscape we know today.

The book is also a monument to the power of closely reasoned argument – built from well organised and carefully evidenced points that are not merely persuasive, but designed to be irrefutable. Indeed, it had to be: upon its publication, the chemical industry utilised all its resources to attempt to discredit both *Silent Spring* and Carson herself – to no avail.

The central argument of the book is that the indiscriminate use of pesticides encouraged by post-war advances in agriculture and chemistry was deeply harmful to plants, animals and the whole environment, with devastating effects that went far beyond protecting crops. At the time, the argument directly contradicted government policy and scientific orthodoxy – and many studies that corroborated Carson's views were deliberately suppressed by hostile business interests. Carson, however, gathered, organised and set out the evidence in *Silent Spring* in a way that proved her contentions without a doubt.

While environmental battles still rage, few now deny the strength and persuasiveness of her reasoning.

ABOUT THE AUTHOR OF THE ORIGINAL WORK

Rachel Carson was born in 1907 and spent her childhood on the family farm in Springdale, Pennsylvania, in the United States. Her rural upbringing laid the foundations for a deep connection with the natural world, and this would emerge later when Carson became one of the earliest environmentalist writers and campaigners. As an academic, Carson studied biology and later zoology at the Pennsylvania College for Women, Woods Hole Institute, and Johns Hopkins University. She then held scientific positions with the US government for 15 years. Carson devoted her later years to writing and produced four best sellers on environmental themes. She died in 1964 of breast cancer, aged just 56.

ABOUT THE AUTHOR OF THE ANALYSIS

Nikki Springer has studied at MIT, Harvard and Yale. She is currently researching her PhD in environmental management at Yale.

ABOUT MACAT

GREAT WORKS FOR CRITICAL THINKING

Macat is focused on making the ideas of the world's great thinkers accessible and comprehensible to everybody, everywhere, in ways that promote the development of enhanced critical thinking skills.

It works with leading academics from the world's top universities to produce new analyses that focus on the ideas and the impact of the most influential works ever written across a wide variety of academic disciplines. Each of the works that sit at the heart of its growing library is an enduring example of great thinking. But by setting them in context – and looking at the influences that shaped their authors, as well as the responses they provoked – Macat encourages readers to look at these classics and game-changers with fresh eyes. Readers learn to think, engage and challenge their ideas, rather than simply accepting them.

WAYS IN TO THE TEXT

KEY POINTS

- Rachel Carson was one of the most influential environmental* writers of the twentieth century.

- Her book *Silent Spring* is widely credited as sparking the environmental movement* in the United States.

- More than 50 years after its publication in 1962, *Silent Spring* remains a foundational text for environmental students and professionals.

Who Was Rachel Carson?

Rachel Carson (1907–64) was an American environmental writer whose greatly influential book *Silent Spring* (1962) is commonly credited with launching the environmental movement (a political and social movement emphasizing the importance of the conservation of the natural world, respecting the finite nature of our natural resources, and so on) in America in the 1960s and 1970s.

Carson's mother, a strong influence, encouraged her to develop a deep connection, respect, and love for the natural world. After a rural upbringing on the family farm in the state of Pennsylvania, she won a scholarship to the Pennsylvania College for Women, where she initially majored in English. She changed courses early in her university career, however, and graduated in 1929 with a degree in biology*—the study of living organisms. She continued her studies at the Woods Hole Institute and Johns Hopkins University, where she was awarded an MA in zoology* (the study of animal life) in 1932.[1] Carson worked as a scientist and writer for the US government for 15 years before she became an independent full-time writer.

Carson's writings, and *Silent Spring* in particular, were not only central to the launch of the environmental movement, but they had a particular influence on public awareness about the need for a more stringent regulation of pesticides*—chemicals used to control pests, notably insects,* that threaten the health of cultivated plants. The agriculture and pesticide industries attacked *Silent Spring's* ideas furiously. In response, Carson pointed out how political loopholes allowed these industries to operate with little regard for the environmental consequences of their products and methods. She was not able to continue this work for long, however; Carson died of breast cancer in 1964 aged just 56.

What Does *Silent Spring* Say?
Silent Spring is a cautionary tale about the world and its ecosystems*— its various systems of habitats, plants, and animals—for anyone who wishes to be a thoughtful, knowledgeable citizen of the earth. It has helped millions of academics, policymakers, and concerned citizens recognize the impact of humans on nature and consider how to safeguard current and future generations by finding a sustainable way of life.

The main subject of Carson's *Silent Spring* is the negative, widespread, and long-lasting effects of human activity on the environment, which she illustrates through one major case study: the use of chemical pesticides in agriculture. Carson details the harmful and devastating results of the widespread use of synthetic chemical pesticides by farmers in the United States after World War II* (1939– 45). These include everything from cancer* and birth defects in humans to the killing of small animals and birds. In fact, the "silent" of the title refers to the lack of singing birds after pesticides have destroyed the world's wildlife. Carson called for such chemicals to be renamed "biocides"*—a more accurate reflection, she believed, of their true strength and danger. The warnings of *Silent Spring* are generally credited with leading to the banning of DDT,* the pesticide dichloro-diphenyl-trichloroethane, in the United States in 1972.

Carson notes the irony that the United States government subsidized farmers to produce a surplus while claiming that chemical pesticides were needed to ensure an adequate food supply. She also criticized the political system for creating a situation that favored the claims of industry giants over the research of independent scientists and doctors into the safety of chemicals and their effects on the health of both humans and wildlife.

Silent Spring ultimately advocates the use of biological pest control.* Carson suggests using the natural checks and balances of ecosystems to control pests—for instance, diversifying crops (expanding the varieties grown) rather than planting monocultures* (growing a single species over a wide area). This, she believed, would ensure the security of the food chain while preventing damage from synthetic pesticides.* Carson's ideas about a more environmentally friendly manner of controlling agricultural insects was important, as it was a direct denial of the claims made by the chemical industries that pollution was an unavoidable consequence of an abundant and profitable food supply. She also called for a radically different approach to the use of pesticides, from one she deemed "indiscriminate" to one that carefully limits the amount and range of application.

Underpinning Carson's research on the harmful impact of chemical pesticides is a larger question: Should we try to control the environment and its ecosystems for our own, often short-term, purposes, regardless of the long-term impact on wildlife, nature, the health of the planet, or the safety and welfare of future generations?

Why Does *Silent Spring* Matter?

Silent Spring is often cited as the most important piece of modern environmental literature. It was pivotal in alerting the public to the dangers of pesticide use and to the risk of releasing into the environment chemicals whose long-term effects were not fully, or even marginally, understood. It had an equally significant impact within the scientific community.[2]

The publication of *Silent Spring* changed environmental regulation in the United States. In addition to the ultimate ban on DDT in agriculture, Carson's influence was also evident in the creation in 1970 of the Environmental Protection Agency (EPA).* This helped to separate the regulation of pesticides and chemicals from agricultural subsidies, including within the large-scale industrial operations commonly referred to as "big agro." Before this, both functions had been handled by the United States Department of Agriculture, but Carson considered this a conflict of interest.

Carson and her work continue to influence current-day policies and politicians. In the 1994 preface to *Silent Spring* written by Al Gore,* who was vice president of the United States at the time, Gore remarks that a portrait of Carson hung in his office to serve as a constant reminder of her work and its importance.[3]

Carson's impact reached far beyond the use of pesticides. It called into question the trustworthiness of the relationships between major corporations and government regulators, and the credibility of the sweeping claims often presented to the public as factual information but unsubstantiated by independent research. For many in the comparative comfort of postwar society, the realities presented in *Silent Spring* came as a shock. The dangers the book presented were a far cry from the marketing images of happy children frolicking on a perfectly manicured lawn found on the front of the fertilizer bag purchased at the local hardware store. If Carson's claims were true, what other horrors lay in plain sight?

Carson also helped to pave the way for other female scientists and is a forerunner of the ecofeminism* movement, which began shortly after *Silent Spring's* publication. Ecofeminism is a political movement in which issues of environmental conservation* are combined with the principles informing the struggle for equality between the sexes. Carson remains a role model today for anyone concerned about the environment and is a constant reminder of the power of a single voice against even the most powerful of interests.

NOTES

1   Linda Lear, *Rachel Carson: Witness for Nature* (New York: Macmillan, 1998), 72.

2   Joshua Rothman, "Rachel Carson's Natural Histories," *New Yorker*, September 27, 2012.

3   Al Gore, introduction to *Silent Spring*, by Rachel Carson (New York: Houghton Mifflin, 1994), xviii.

SECTION 1
INFLUENCES

MODULE 1
THE AUTHOR AND THE
HISTORICAL CONTEXT

KEY POINTS

- Rachel Carson was deeply influenced by her rural upbringing on a family farm, which led to her profound understanding of the natural world.

- Carson's studies of both English and biology* enabled her to communicate complex scientific ideas clearly to a wide general readership.

- *Silent Spring* was one of the first pieces of environmental literature to be embraced by politicians and the general public when it was published in the United States in 1962.

Why Read This Text?

In the field of environmental literature, Rachel Carson is one of the most notable and quoted authors of the twentieth century; her book *Silent Spring* (1962) is required reading for all students of ecology* (the branch of biology that focuses on the relationship between groups of living things and their environment), whether or not they agree with Carson's conclusions.

Although most of the scientific data and research presented in the book concentrates on the effects of agricultural pesticides,* its themes and lessons can be applied much more generally. Indeed, *Silent Spring* is just as relevant today as it was when it was published more than 50 years ago. As a result, the number of books and scholarly articles written directly about Carson and *Silent Spring* continues to rise.

Silent Spring also remains a prime example of the power of one person, and one voice, standing against a worldwide problem. Carson,

> ❝ We must not be selfish or timid if we hope to have a decent world for our children and grandchildren. ❞
> Former United States President Jimmy Carter, televised speech, April 18, 1977

who died soon after *Silent Spring*'s publication, continues to show how an "everyday" person can improve society and the planet. She was a pioneer in the field of modern environmental literature and a catalyst of the modern environmental movement* in the United States and elsewhere.

Author's Life

Rachel Carson (1907–64) had a lifelong relationship with the natural world, nurtured during her childhood on her family's 65-acre farm in the district of Springdale, Pennsylvania. She was an avid reader and a skilled writer, and had her first story published in the children's magazine *St. Nicholas* at the age of 11. Most of her reading and writing focused on themes from nature and she had a particular interest in the oceans, a topic that she would write about professionally later in life.[1]

Carson excelled academically and graduated top of her high school class in 1925 before enrolling in the Pennsylvania College for Women (now called Chatham University) in Pittsburgh. She began by studying English, but soon switched to biology, though she continued to write for the university's newspaper and literary magazine. Carson completed her undergraduate program *magna cum laude* ("with great distinction") in 1929, and went on to take a master's in zoology* in 1932 at Johns Hopkins University in Baltimore, Maryland.[2]

She had begun work at Johns Hopkins on a doctoral degree, but was forced to leave in 1934 and find a job for financial reasons. Soon after, Carson's father died, and Rachel took on the full financial responsibility of caring for her mother. At the urging of a college mentor, she applied for a temporary position at the US Bureau of Fisheries* (now the US Fish and Wildlife Service, the main conservation agency within the Department of the Interior), where her primary responsibility was writing copy for radio programs. She also began to write a regular column on natural history topics for the *Baltimore Sun* newspaper. Her first notable professional publication was in July 1937, when her essay "Undersea," which beautifully described the wonders of the ocean floor, appeared in the *Atlantic Monthly*, a major American magazine. In fact, the essay was so impressive that Carson received invitations from publishers to expand it into a book.[3]

Carson had a successful career and became chief editor of publications for the US Bureau of Fisheries by 1949. She continued, though, to write and publish independently. Her first three full-length books, *Under the Seawind* (1941), *The Sea Around Us* (1951), and *The Edge of the Sea* (1955), all explored her interest in nature generally and in the oceans and aquatic life in particular.

Author's Background

The most important early influences on Carson's life are all rooted in her rural upbringing on the family farm, which gave her a fundamental love of and respect for the natural world. She understood that nature operated in mysterious, complex ways, and that it was vital to try to understand the natural world on its own terms, rather than attempt to dominate and control it for human use.

What started as a temporary position at the US Bureau of Fisheries, taken to support her family when bereavement forced her to quit the academic world, turned into a 15-year government career for Carson. The success of the books she published on environmental topics—*The Sea Around Us* and *The Edge of the Sea*—provided her with the financial means to resign from government work and devote her time and attention fully to writing.

NOTES

1   Linda Lear, *Rachel Carson: Witness for Nature* (New York: Macmillan, 1998), 120.

2   Lear, *Rachel Carson*, 63.

3   Lear, *Rachel Carson*, 88.

MODULE 2
ACADEMIC CONTEXT

KEY POINTS

- In *Silent Spring*, Carson wrote about the dangers of widespread chemical pesticide* use in the United States.

- Building on the emerging field of environmental* literature, *Silent Spring* is considered, along with the author and environmentalist Aldo Leopold's* *Sand County Almanac*, to form the foundation of twentieth-century environmental literature.

- Carson drew upon both her rural background and her scientific training in writing the book.

The Work in its Context

Rachel Carson's *Silent Spring* was published during a period of tremendous scientific advancement in the years following World War II.* Government and industry wholeheartedly embraced this idea of progress and fostered a culture that celebrated technology, science, and modernization. Cleanliness, standardization, technological development, quality control, and industrial mass production were all seen as signs of social progress. Human advancement and control of the natural world were interconnected themes and they were reflected in everything from putting a man on the moon to the popularization of processed, compartmentalized TV dinners.

These cultural values of scientific precision and highly technical feats of engineering also affected society's attitude toward the natural world, resulting in highly maintained, aesthetically specific landscapes (landscapes cultivated and shaped according to certain ideas of what is beautiful), the elimination of pests, and the maximization of efficiency

❝ Even if she had not inspired a generation of activists, Carson would prevail as one of the greatest nature writers in American letters. ❞

Peter Matthiessen, "100 Most Influential People of the Century," *Time* magazine

in agricultural production. "Science and technology and those who worked in these fields were revered as the saviors of the free world and the trustees of prosperity," said Carson's biographer Linda Lear.[1]

One of the reasons behind this state of mind was the civilian use of military technology and chemicals developed during World War II, including chemical pesticides, as well as the medical advances related to the prevention and cure of insect-borne illnesses.* Another contributory factor was the rise of huge corporations in American business and industrial control over regulation (that is, control by industry of restrictions set by the government on business practices). Corporations were seen as wise, trustworthy, scientifically advanced entities that were the source of much of the postwar economic growth of the United States in the 1950s.

Overview of the Field

The modern environmental movement* was born out of the massive increase in pollution started by the Industrial Revolution,* a period of intense technological and industrial growth that began in Britain in the late eighteenth century, and in the course of which Western societies turned from agricultural to industrial economies. This growth in pollution continued unchecked into the twentieth century. While technological advances allowed for unparalleled economic growth and provided many new jobs for working-class people, the resulting environmental and health impacts soon became apparent in deteriorating air quality, water quality, and overall environmental quality. It was also clear that human health suffered.

The conservation movement* in the United States brought about a political and cultural focus on the natural world and its preservation and conservation. It lasted from around 1890 to 1920 and built on the work of figures such as the writer and political activist Henry David Thoreau,* the poet Ralph Waldo Emerson,* and the Scottish environmental philosopher John Muir.*[2] It was popularized, too, by the United States President Theodore "Teddy" Roosevelt,* who served in office from 1901 until 1909. He championed the creation of the United States Forest Service (an organization concerned with the management of the nation's forests), signed into law the Antiquities Act, which gave the president the authority to create National Monuments (places considered significant for their special cultural or natural importance), and created five National Parks—land protected from exploitation on account of its particularly unique qualities.[3] The two World Wars of 1914–18 and 1939–45 and the catastrophic economic downturn of the 1920s and 1930s known as the Great Depression, however, shifted focus away from the environment. It was not until the late 1950s and 1960s that environmental leaders started to worry about factors such as economic growth, suburban development, paternalistic corporations (business organizations prescribing solutions and behavior for the nation's citizens), modernization and mechanization, and their effects on the natural world. This was certainly due in part to Carson's work.

In the 1950s, when Carson was working on *Silent Spring*, there was little environmental literature around. One important exception to this was the author and environmentalist Aldo Leopold's *Sand County Almanac: And Sketches Here and There*, published in 1949. In it, Leopold coined the term and developed the idea of a "land ethic,"* a philosophy that describes man's responsibility for a respectful relationship with nature and the landscape. *Silent Spring* itself would become the inspiration for many of the federal environmental regulations implemented during the administration of President Richard Nixon

(1969–74), such as the Clean Water Act (1972),* the Clean Air Act (1970),* and the National Environmental Protection Act (1970)*— all laws passed to protect the nation's health and environment.

Academic Influences

The research ultimately presented as *Silent Spring* developed from Carson's other work on chemical pesticides. In the post–World War II era, the United States military funded a number of research and development programs for the agricultural application of chemical pesticides. One such program was the gypsy moth* eradication program in the mid-Atlantic region, which involved widespread, indiscriminate spraying of the pesticide DDT* on large swathes of land. Infestations of the gypsy moth, an insect* introduced from Europe in the nineteenth century whose larvae eat the leaves of plants, had caused widespread defoliation in forested areas; this had a measurable impact on the health and behavior of birds, which were forced to moved from now leafless ("defoliated")* trees.[4]

Carson's focus on pesticides—and specifically DDT—began when she was contacted by private landowners on Long Island, New York, concerned about this indiscriminate DDT spraying on their lands by the federal government. A 1958 letter to the editor of the *Boston Herald* caught Carson's attention. It outlined the avian (bird) fatalities resulting from DDT spraying to eradicate mosquitos. Carson was quite knowledgeable already on this subject but she was encouraged to research it further by such groups as the Washington,[5] DC chapter of the Audubon Society,* a major conservation advocacy group that largely supported the crusade against chemical pesticides. It commissioned Carson to study the government's spraying programs and its consequences to help it publicize the potential dangers involved.

Carson received guidance from practicing and academic scientists, who reviewed many of the technical details in *Silent Spring*. She conducted much of her research at the National Institutes of Health* Medical Library, and worked with many of the research fellows there. The most significant guidance came from Dr. Wilhelm Hueper,* a leader in the identification of cancer*-causing pesticides, a subject which at that time remained a questionable and controversial theory. Carson also drew on support from the numerous government scientists she had worked with during her tenure at the US Bureau of Fisheries.* She soon realized that the academic and scientific community was divided concerning its views on the impact of chemical pesticides. It was a difference of opinion that would only intensify after *Silent Spring* was published.[6]

NOTES

1   Robin McKie, "Rachel Carson and the Legacy of Silent Spring," *Guardian*, May 26, 2012.

2   Max Oelschlaeger, "Emerson, Thoreau, and the Hudson River School," *Nature Transformed*, National Humanities Center. Accessed December 30, 2015, http://nationalhumanitiescenter.org/tserve/nattrans/ntwilderness/essays/preserva.htm.

3   Douglas Brinkley, *The Wilderness Warrior: Theodore Roosevelt and the Crusade for America* (New York: Harper Perennial, 2010), 76–80.

4   Laurent Belsie, "Gypsy Moths Return to Northeast: Worst Outbreak in a Decade Descends on Northeast; Entomologists Do Not Know How to Stop It," *Christian Science Monitor*, July 2, 1990.

5   Patricia H. Hynes, *"Unfinished Business: Silent Spring on the 30th anniversary of Rachel Carson's Indictment of DDT, Pesticides Still Threaten Human Life,"* Los Angeles Times, September 10, 1992.

6   Bryan Walsh, "How *Silent Spring* Became the First Shot in the War on the Environment," *Time*, September 25, 2012.

MODULE 3
THE PROBLEM

KEY POINTS

- *Silent Spring* is a book about the harmful impacts to plants, animals, and humans caused by chemical pesticides* used to control insects and protect agricultural production.

- One of Carson's most significant partners was Dr. William Hueper,* a pioneer in the field of environmental cancer.*

- Carson took a passionate, moralistic stance on the need for individuals and policymakers to keep big business in check and ensure protection of the natural world.

Core Question

At its core, Rachel Carson's *Silent Spring* is a book about the harmful effects of the widespread and indiscriminate use of chemical pesticides in agricultural production, an idea that was at that time extremely controversial. She also addresses the overuse, and industry encouragement of, domestic pesticide application by homeowners. Much of the book is devoted to bringing together different strands of scientific data that indicate these chemicals' wide variety of detrimental impacts on plants, animals, and people.

The data and impacts presented by Carson are important for several reasons. First, at the time, the use of pesticides was primarily viewed as technological progress and a necessary tool to enable farmers to increase their yields while preventing the spread of insect-borne illnesses.* Second, the government supported the use of pesticides and worked in conjunction with large agricultural corporations to tell the American public with one voice that pesticides were safe and

> ❝ The 'control of nature' is a phrase conceived in arrogance, born of the Neanderthal age of biology* and philosophy, when it was supposed that nature exists for the convenience of man. The concepts and practices of applied entomology [the study of insects]* for the most part date from that Stone Age of science. It is our alarming misfortune that so primitive a science has armed itself with the most modem and terrible weapons, and that in turning them against the insects it has also turned them against the earth. ❞
>
> Rachel Carson, *Silent Spring*

beneficial. The government was also responsible for the indiscriminate use of pesticides on public land. Third, the presentation of Carson's research would call into question the trustworthiness both of large, paternalistic corporations—corporations that dictated things such as environmental* solutions to their customers—and of the government itself for not putting the health, safety, and welfare of its citizens first. Finally, *Silent Spring* was the catalyst for much of the environmental regulation that followed in the coming decades.

The Participants

Carson was well aware that the publication of *Silent Spring* would cause an uproar.[1] It is a serious critique of the power and trust given to large corporations and of the inevitable conflict of interest for government agencies that support and subsidize farmers while simultaneously regulating food and agricultural production. In the 1950s, American corporations were often considered benevolent. The public placed a good deal of trust in them, particularly those in the chemical, automotive, and other technological fields, as they were considered to be key to the post-World War II* economic prosperity enjoyed in the United States.[2]

Carson was not alone, however, in her quest to reveal the truth behind modern chemical use. Scientists at the National Institutes of Health (NIH),* a federal agency conducting medical research, had already been collecting data on the impact of chemical pesticides for several years, and some major environmental advocacy groups, including the Audubon Society,* had also been mounting campaigns to educate the public on the issue.[3] Carson is seen by many as the unifying force who brought these concerns to light, capitalizing on her proven literary elegance to make an indelible impression on the public mind.

Of particular importance to Carson's research was Dr. Wilhelm Hueper, a scientist with the National Cancer Institute.* Hueper was one of the leaders of the environmental cancer laboratory and his work linked several types of pesticides to cancer in humans and animals. In 1942, Hueper published *Occupational Tumors and Allied Diseases*, one of the first medical books to directly link occupational hazards to cancer and label certain industrial substances as carcinogens* (substances capable of causing cancer).[4] Dr. Hueper's previous experience as a scientist for research laboratories funded by the DuPont Corporation, a major chemical company, was of particular interest to Carson. He had left his DuPont job after publishing data on the harmful impacts of industrial chemicals on DuPont's own employees.[5]

Ultimately, Carson found the collective scientific community to be divided. On the one hand, some supported Carson's opinion that chemical and synthetic pesticides* had a measurable impact of the health of humans, wildlife, and ecosystems.* These scientists had been keeping much of their work hidden from public view, however, fearing that it would be highly controversial to contest government-supported programs and large corporate interests. On the other hand, Carson was dismayed to find that a number of scientists simply dismissed the dangers of pesticide use. Having worked as a government employee, though, she was aware that there were could be serious economic repercussions from any data that suggested pesticides were anything other than harmless.[6]

The Contemporary Debate

Carson's work continues to be regarded as one of the most influential pieces of modern environmental literature and a constant reminder of the danger of taking action without understanding the consequences. This idea is now popularly known as the precautionary principle.* Carson's work, together with Aldo Leopold's* *Sand County Almanac* (a work calling for a more respectful and thoughtful relationship between human beings and the land they inhabit), forms the literary basis of the environmental and conservation movement* today.

Carson's work was not only important to the foundation of environmental literature, but it also helped stimulate the public activism that led to many of the laws, regulations, and policies that now form the field of environmental law and policy in the United States.[7] *Silent Spring* gave the American public permission to question the ecological and public health impacts of technological and economic advancement and instilled a sense of responsibility in people to keep such impacts in check. Such attitudes are now commonly referred to as "earth stewardship."*

Carson paved the way for many other environmental and social activists, particularly female ones, including the anthropologist Margaret Mead* and Erin Brockovich,* a law clerk who was primarily responsible for one of the largest class-action lawsuits, brought against the Pacific Gas and Electric Company in San Francisco in 1993 for its contamination of drinking water.[8] Many believe the worldwide celebration of Earth Day,* begun in 1970 and held every year on April 22, is directly related to Carson's groundbreaking work and fearless environmental leadership, and many Earth Day celebrations pay tribute to her work.[9]

NOTES

1   Linda Lear, *Rachel Carson: Witness for Nature* (New York: Macmillan, 1998), 446–8.

2   See the work of Charles and Ray Eames, for example, and their marketing of the IBM Corporation. Eric Schuldenfrei, *The Films of Charles and Ray Eames: A Universal Sense of Expectation* (New York: Routledge, 2015).

3   Vera L. Norwood, "The Nature of Knowing: Rachel Carson and the American Environment," *Signs* (1987): 740–60.

4   Wilhelm C. Hueper, "Occupational Tumors and Allied Diseases," *Occupational Tumors and Allied Diseases* (1942).

5   Devra Davis, *The Secret History of the War on Cancer* (New York: Basic Books, 2007), 77.

6   Lear, *Rachel Carson*, 330–5.

7   Gary Kroll, "The 'Silent Springs' of Rachel Carson: Mass Media and the Origins of Modern Environmentalism," *Public Understanding of Science* 10, no. 4 (2001): 403–20.

8   Scott Gillam, *Rachel Carson: Pioneer of Environmentalism* (Edina, MN: ABDO, 2011), 91.

9   See, for example, the Earth Day Network or the PBS series *American Experience*.

MODULE 4
THE AUTHOR'S CONTRIBUTION

KEY POINTS

- In *Silent Spring*, Carson argues that the widespread use of chemical pesticides* is placing both the natural balance of the world and human beings in danger.

- Carson combined a descriptive, poetic writing style with hard scientific data to present her claims in a way that was both alarming and accessible to the public.

- Although Carson's claims about pesticides were not exclusively her own, she communicated the data and issues in a direct and lucid way that brought her a wide readership.

Author's Aims

As a trained scientist and long-time lover of nature, Rachel Carson was objectively aware of the dangers of indiscriminate pesticide use; *Silent Spring* was her means of announcing this to the world. Her extensive experience as an environmental* writer, combined with her in-depth training as an academic and professional scientist, made her uniquely able to communicate this message. Carson's primary aims in *Silent Spring* were to reveal how the government and industry were indiscriminately spraying pesticides on public and private lands; and to alert the world to the dangers of chemical pesticide use and exposure.

Carson makes several powerful, controversial arguments. First, she outlines specific data that she and her scientific colleagues have collected about the widespread use of chemical and synthetic pesticides* and the dangers they present to plant, animal, and human life. Second, she calls into question the inherent conflict of interest in

> 66 [*Silent Spring*] continues to be a voice of reason breaking in on our complacency... [Carson] brought us back to a fundamental idea lost to an amazing degree in modern civilization: the interconnection of human beings and the natural environment. 99
>
> Al Gore,* Introduction to *Silent Spring*

the organization of the United States federal government. On the one hand, it controls the testing and regulation of toxic (or potentially toxic) substances, but on the other, it also runs the programs that support farmers, the agricultural industry, and food regulation. Third, she presents her most important argument—that the short-term gains afforded by technological advancement, particularly those related to chemical use, are liable to cause detrimental, long-term effects that more than outweigh the benefits.

Carson was not alone in her aims. In fact, other scientists and environmental conservation* groups recruited her to join their work, to make use of her ability to craft beautiful prose that would speak to the heart as well as to the mind in a way that was far more effective than bald scientific data. Carson also paved the way for countless other environmental activists who, while focused on different topics, imitated her approach in the hope of matching her success.

Approach

Carson's words present much more than raw data and scientific fact. They paint an elegant picture of nature in balance, and contrast this with horrifying examples of the impact of synthetic pesticides. This brought her a wide-ranging readership that included scientists, politicians, and the general public. The author Mark Hamilton Lytle* called her a "gentle subversive" in his 1998 book of the same name.[1] In it, he chronicles the way she delicately balanced her ladylike, demure public presence with her aggressive challenge to some of the greatest powers in society.

Silent Spring's first chapter, "A Fable for Tomorrow," presents a haunting account of the environmental reality that would follow from the continued use of pesticides: "Then a strange blight crept over the area ... some evil spell had settled on the community ... everywhere was a shadow of death ... there was a strange stillness. The birds, for example—where had they gone?" Carson goes on to outline a number of devastating and disturbing changes in plant life, wildlife, and humans, but concludes by reassuring the reader that "no community has experienced all the misfortunes I describe ..." She cautions, though, that "every one of these disasters has actually happened somewhere, and many real communities have already suffered a substantial number of them."

Contribution in Context

Carson relied heavily on the research of a small community of scientists whom she met primarily during her work at the US Bureau of Fisheries.* Many of them were government employees working at the National Institutes of Health (NIH)* and the National Cancer Institute (NCI).* These scientists were already documenting and analyzing the harmful effects of synthetic pesticides and, in particular, classifying them as carcinogens.* It was a conclusion that was particularly controversial at the time and many of them would have kept their research silent were it not for Carson.

Carson also relied on the approach she had perfected in her previous highly acclaimed books. She kept the public engaged with her elegant and fluid language, carefully mixing in scientific data and her own original research. Carson's research partners were sometimes dismayed at her efforts to include "softer" language. However, for Carson these two sides to her approach—a literary style that captured her amazement and wonder of the natural world, and the scientific rigor of her training that allowed her to understand it—were inseparable. This approach helps Carson remain remembered and relevant today.

NOTES

1   Mark Hamilton Lytle, *The Gentle Subversive: Rachel Carson, Silent Spring, and the Rise of the Environmental Movement* (New York, Oxford: Oxford University Press, 2007).

SECTION 2
IDEAS

MODULE 5
MAIN IDEAS

KEY POINTS

- In *Silent Spring*, Carson examines the impact of indiscriminate pesticide* use and the harmful effects it has on plant, animal, and human life.

- The central idea of *Silent Spring* is that man's technological power and presumption of control over nature has dire and long-term consequences.

- *Silent Spring* presents its ideas in a contrasting combination of scientific data and cautionary storytelling, which gives the overall message both objective validity and emotional connection.

Key Themes

In *Silent Spring*, Rachel Carson alerts the public and policymakers alike to the risks of chemical and synthetic pesticide* use. More broadly, she warns of the potential dangers if society commits to technological advancements leading to human control over nature without understanding the consequences. Carson shows evidence that chemicals remain in the environment* for long periods of time, and that their harmful impacts can be passed from species to species—a process called bioaccumulation*—and even from mother to child.

The work's three main themes are: the dangers and toxicity of chemical and synthetic pesticides; how such chemicals are recklessly used in the environment; and the alternatives to pesticides that pose a lesser threat to the environment. *Silent Spring* introduces the idea that earth stewardship* is a moral obligation for all human beings. Regarding the killing of innocent animals from exposure to pesticides,

> **❝** Every once in a while in the history of mankind, a book has appeared which has substantially altered the course of history. **❞**
>
> United States Senator Ernest Gruening, in Julia Keller, "Works such as *Uncle Tom's Cabin* and *Silent Spring* had such a Profound Political Effect that they Became ... Art that Changed the World"

Carson asks, "By acquiescing in an act that causes such suffering to a living creature, who among us is not diminished as a human being?"[1]

Carson's conclusion is a recommendation that government policy and agricultural practice be changed to make use of the inherent defenses nature provides when ecosystems* are healthy, robust, and resilient. She offers specific strategies such as natural pesticides* (nonchemical means of pest control), crop rotation* (the practice of regularly changing what is cultivated in a particular place), and biological pest control* (the practice of managing landscape and agricultural pests through the use of natural biological controls such as parasites, predatory insects,* and so on).

Exploring the Ideas

A good portion of *Silent Spring* presents research and data that support Carson's claim that the use of DDT* (the chemical dichloro-diphenyl-trichloroethane) as a chemical pesticide causes significant harm to plants, animals, and people. In particular, she says, the public are being continually exposed to this and other chemicals at alarming rates and without their knowledge or consent. She organizes the impacts of pesticide use into separate chapters based on individual resource systems: water, soil, plants, animals, and people.

DDT is a man-made organic compound first created in 1874. Its insecticidal properties were discovered in 1939 by the Swiss chemist Paul Müller,* winning him the Nobel Prize in 1948. The US Military began using DDT to control head lice in troops during World War II,* and a surplus of DDT after the war, in addition to the fact that air force pilots were now home from combat, contributed to the federal government's decision to use it, sprayed from planes, to control and eradicate certain insect species. These included the gypsy moth* (a pest introduced from Europe in the nineteenth century that threatens native plant life) and the fire ant (an insect imported into the United States from elsewhere in the Americas in the 1930s). At the same time, the government authorized heavy use of pesticides by the agricultural industry.

Carson and her peer scientists had documented numerous harmful side effects from the spraying of DDT. These included the thinning of eggshells laid by multiple varieties of birds—and thinner eggshells lead to a significant decrease in the rate of survival of young birds. Carson's title *Silent Spring* refers to a silent future spring without birds due to the effects of DDT and other chemical pesticides. Carson also provided data on other detrimental side effects of DDT, including cancer,* endocrine disruption* (the disruption of our capacity to secrete hormones into the bloodstream), and premature birth. In addition, as it moves up the food chain, the concentration of DDT increases in organisms through a process called bioaccumulation, which increases the toxicity of the chemical for larger and more sophisticated species such as large mammals* and eventually humans.

Carson also details the widespread use of DDT in the chapter "Needless Havoc," writing that "we are adding … a new kind of havoc—the direct killing of … practically every form of wildlife by chemical insecticides indiscriminately sprayed on the land … the incidental victims of [man's] crusade count as nothing."[2] Carson cites examples of this spraying. In some cases, the targeted pest was not considered to be of significant concern or population to be worth further investigation. In others, the chemicals used had not been shown to be particularly effective against the species or safe for other species sharing the same habitat.[3]

One of the most radical elements in *Silent Spring* is Carson's comparison of DDT's impact on the environment to that of radiation* from the atom bomb,* the memory of which was still prominent in the minds of the American public less than 20 years after the US military dropped atom bombs on the Japanese cities of Nagasaki and Hiroshima, killing well over 100,000 civilians. She says in *Silent Spring*, "We are rightly appalled by the genetic effects of radiation. How then, can we be indifferent to the same effect in chemicals that we disseminate widely in our environment?"[4]

Language and Expression

The uniqueness and power of *Silent Spring* is directly related to its poetic elegance, Carson's facility with words, and her ability to paint a picture with language. She is able to take a set of very technical data and communicate it to a large and diverse audience, and most importantly, help them understand why we need to be concerned.

Carson was already a well-known, award-winning author before she published *Silent Spring*, winning the National Book Award for Nonfiction in 1952 for *The Sea Around Us*. She uses numerous literary devices, particularly cautionary metaphors, with chapter titles such as "Elixirs of Death," "Rumblings of an Avalanche," and "The Other Road." The text is full of (equally cautionary) storytelling. These devices help her to reach readers on an emotional level and help them personally connect with the scientific issues. For many readers, this emotional connection generates a sense of obligation and responsibility and fuels a strong response to her "call to action."

Carson understood her audience well and played to their fears and self-interests. For instance, to strike a chord with suburban housewives, she described in detail the sight of common animals, such as raccoons and squirrels, discovered dead and disfigured on their pristine manicured lawns. This is an example of her knack of communicating with readers who would have an emotional as well as rational response to her ideas. In the 1960s, women were not afforded the same respect and scientific acknowledgement as men, but women were going to be primarily concerned for their own small children playing in the backyard.

NOTES

1   Rachel Carson, *Silent Spring* (New York: Houghton Mifflin Harcourt, 2002), 100.

2   Carson, *Silent Spring*, 85.

3   Carson, *Silent Spring*, 85–100.

4   Carson, *Silent Spring*, 37.

MODULE 6
SECONDARY IDEAS

KEY POINTS

- The main secondary idea in *Silent Spring* is that everyone has a moral responsibility to protect the environment.*

- Carson highlights the conflicts between environmental protection and economic profit.

- *Silent Spring* is also a general call to action, encouraging everyone to demand information from government and educate themselves about these issues.

Other Ideas

In *Silent Spring*, Rachel Carson demands that the American public consider their moral obligations. They have, she says, a duty to protect nature and help those unable to help themselves—namely, wildlife, children, and future generations who will otherwise inhabit a contaminated world: "Future generations are unlikely to condone our lack of prudent concern for the integrity of the natural world that supports all life."[1]

Underlying Carson's *Silent Spring* are ideas that are common to most discussions on environmental policy and regulation, such as the inherent conflict between economics and business on the one hand, and environmental protection on the other. Carson's stance is that the natural world deserves more respect than corporate profits. However, she also uses economic arguments to her advantage. Not only can biological controls* cost less per acre, she says, but there are further savings when there are no poisoned wildlife or food products to deal with and no effects on human health.[2]

❝ The sedge is wither'd from the lake, And no birds sing. ❞

John Keats, "La Belle Dame Sans Merci"

Carson devotes an entire chapter to the relationship between pesticides,* genetics* (the science of genes, the biological material allowing the transmission of characteristics from generation to generation), and cancer.* This idea was quite new at the time of publication, and considered controversial given the economic and political power of major chemical corporations. Several of Carson's scientific colleagues, most notably Dr. Wilhelm Hueper,* continued to have long and notable careers in environmental cancer research. Carson also highlights the inherent bureaucratic conflicts of interest within the federal government. These, she argues, contribute to the widespread and indiscriminate use of the pesticide DDT* without public knowledge—an idea she would elaborate on in her post-publication Senate hearings.

Exploring the Ideas

One of the most significant compliments Carson received was having *Silent Spring* compared to the nineteenth-century author Harriet Beecher Stowe's* *Uncle Tom's Cabin*. Stowe's book, published in 1852, helped galvanize the abolitionist movement in America—the movement founded in order to outlaw the institution of slavery—and made the case for the American public to consider slavery as a moral sin. It was one of the most popular and controversial books of the nineteenth century. *Silent Spring*, according to the US government's Environmental Protection Agency (EPA),* "played in the history of environmentalism roughly the same role that *Uncle Tom's Cabin* played in the abolitionist movement."[3]

Putting earth stewardship* in a moral context makes it personal; environmental sustainability (the capacity to live within the means of finite natural resources) and stewardship become a moral issue. This idea is gaining popularity today in the context of global climate change* (long-term change in patterns governing the earth's weather and temperature). According to Robb Willer, Professor of Psychology at Stanford University, "People think quite differently when they are morally engaged with an issue. In such cases people are more likely to eschew [reject] a sober cost-benefit analysis, opting instead to take action because it is the right thing to do. Put simply, we're more likely to contribute to a cause when we feel ethically compelled to."[4]

Carson understood the need to first enthrall the public with nature's mysteries and wonders before expecting them to take action to protect it. According to a writer in the *New York Times*, "Carson believed that people would protect only what they loved."[5] Having already established this rapport with readers in her previous books, Carson took this emotional connection just as seriously as she did the use of scientific data and citations. She also introduces a heavy amount of fear and gloom should her warnings not be heeded. Carson explains to her readers the invisible, pervasive, and silent quality of chemical pesticides and their ability to remain stable and potent while lying latent in the environment for many years. She is explicit that government and industry are well aware of the dangers of pesticides and are purposely downplaying or concealing their impacts. No one, she is saying, can trust that the environment or its resources will remain safe.

Overlooked

In addition to its direct commentary on pesticide use, *Silent Spring* was an indirect commentary on postwar, 1950s culture in America—one characterized by consumption (the culture of spending and shopping), materialism (a cultural emphasis on the worth of material goods and the status they confer), new advancements in engineering, science, and technology (notably space exploration), and a new focus on uniformity and factory-produced modernization. Great strides had been made in science and technology in the fields of disease prevention, chemical warfare, and military technology (including the construction of the national highway system in the United States) during World War II.* In the 1950s, many of these advancements were applied to civilian life—and this included the use of the pesticide DDT in domestic cultivation and in the house and garden. In the 1950s, the average citizen was groomed and trained to trust the leadership of the government and large corporations, who, it was believed, had the best interests of the public and the safety and security of America in mind. Carson's demand for public action against government and big business was an early forerunner of the cultural shift toward anti-establishment thought in the 1960s.[6]

Carson's reputation also earned her a place in the fledgling movement of ecofeminism,* a loosely defined term that unites environmental concerns with traditional feminist concerns, seeing both issues as resulting from male domination of society. One can see something of the feminist quality of her approach to nature in one of Carson's most famous statements: "The 'control of nature' is a phrase conceived in arrogance, born of the Neanderthal age of biology* and philosophy, when it was supposed that nature exists for the convenience of man ... It is our alarming misfortune that so primitive a science has armed itself with the most modern and terrible weapons, and that in turning them against the insects* it has also turned them against the earth."[7]

Scholars debate whether ecofeminism is a true subfield of either feminism* (the political and cultural movement associated with the struggle for equality between the sexes) or ecology.* Carson herself, though, was writing at a time when few women were educated in the sciences or had the tenacity to oppose huge, male-dominated industries. She was certainly a pioneer, then, paving the way for other women leaders in the environmental field and elsewhere.

NOTES

1   Rachel Carson, *Silent Spring* (New York: Houghton Mifflin Harcourt, 2002), 15.

2   Carson, *Silent Spring*, 159–72.

3   Joshua Rothman, "Rachel Carson's Natural Histories," *New Yorker*, September 27, 2012.

4   Robb Willer, "Is the Environment a Moral Cause?" *New York Times*, February 27, 2015.

5   Eliza Griswold, "How 'Silent Spring' Ignited the Environmental Movement," *New York Times*, September 21, 2012.

6   Griswold, "How 'Silent Spring.'"

7   Carson, *Silent Spring,* 297.

MODULE 7
ACHIEVEMENT

KEY POINTS

- *Silent Spring* achieved its goal of informing the public about the impacts of pesticide* use.

- Carson's persuasive literary style and her established credibility allowed her to strike a chord among a diverse readership.

- Carson's critics distorted her claims concerning the need for pesticide regulation, attracting other critics to join in their condemnation.

Assessing the Argument

Rachel Carson's *Silent Spring* undoubtedly changed the world, and is largely credited with igniting the environmental movement* in the twentieth century. Carson brought DDT* and its impact on all living things into the public forum. Many credit her book as the single reason why DDT and other chemical pesticides were regulated and banned in the United States, although others believe that they would have naturally fallen out of favor. The website of the US government's Environmental Protection Agency* publicly gives Carson's efforts credit for its own creation, and many environmental leaders and elected politicians give her the overwhelming credit for setting the stage for all major environmental legislation in the United States.

Carson was not, though, without her critics—many of them particularly harsh. The multinational agricultural chemical and biotechnology* company Monsanto* reportedly spent more than $25,000 in 1962 and 1963 to create a public relations case against her and to sing the praises of the chemicals it manufactured and sold.[1]

> **❝** *Silent Spring* fell like a ton of bricks on a wedding party. **❞**
>
> Bill Moyers, *PBS Journal*, 2007

Such campaigns were critical of Carson's work to the point of near-hysteria—labeling her a communist* (a sympathizer of the Soviet Union),* for example, or claiming that speaking against farming in any way is un-American. However, they also no doubt drew attention to her and her book and kept them both in a sensational limelight far longer than they might otherwise have been.[2]

Achievement in Context

Carson's *Silent Spring* had an almost immediate, measurable impact on the United States government. Many hearings in the law-making bodies of Senate and House of Representatives were scheduled in the months after the book's publication and both the president and secretary of the interior commissioned further studies on the pesticides Carson identified. The formation of the Environmental Protection Agency in 1970 is often credited to *Silent Spring*, and several national museum exhibits have been developed and circulated in Carson's honor, including one at the Smithsonian* museum in Washington, DC. The Rachel Carson National Wildlife Refuge near her vacation home in Maine was established in 1969 and is administered by the US Fish and Wildlife Service.*

Carson's ideas reached well beyond environmental, chemical, and political circles. In June 1963, the magazine *Popular Science* published the article "How to Poison Bugs, but NOT Yourself," outlining strategies on less harmful insect* control. In 1962 and 1963, Charles Schultz,* the cartoonist behind the popular *Peanuts* comic strip, referred to Carson on four separate occasions as a "girl's heroine."

Several other comic strips followed suit. In 1969, Joni Mitchell,* a popular female singer-songwriter, paid respect to Carson in the lyrics of her song "Big Yellow Taxi"; the song was rereleased by the group Counting Crows in 2002, evidence of her staying power. Carson's legacy is also honored in many classical and artistic circles, with her connection to the natural world, her superb literary skill, and her unyielding spirit being applauded.[3] Finally, *Silent Spring* has been published in more than 15 languages across Europe, the Americas, and Asia.

Limitations

It can be difficult to assess the relative success or failure of Rachel Carson's *Silent Spring*, as the outcomes of some of her demands are still unclear. For instance, Carson is widely credited with the banning of DDT in the United States in the 1970s, but it continued to be produced for export until as late as 1985, when over 300 tons were exported. Even before the publication of *Silent Spring*, however, production and demand for DDT were beginning to stabilize and later wane as resistant strains of mosquitos were beginning to emerge. Carson forewarned of the development of resistant insect strains, but this was not her own discovery.[4]

Carson called for a renewed focus on biological pest control;* while this is now common in certified organic farms and low-carbon landscapes, it is not widely used. The developed world has certainly been more focused on environmental matters since the publication of *Silent Spring*, but many medical and scientific experts, and even nutritionists (experts in nutrition) and fitness gurus feel that there are still too many synthetic compounds and chemicals in the food we eat, the water we drink, and the cosmetics we apply.

Carson's biographers paint her as a quiet, private woman who did not write *Silent Spring* because she sought the spotlight. She nevertheless remains one of the most popular, influential, and controversial forerunners and advocates of the environmental movement of the twentieth century.

NOTES

1 Rachel Carson Center, accessed December 2, 2015, http://www.environmentandsociety.org/.

2 Rachel Carson Center.

3 Rachel Carson Center.

4 See Rachel Carson, *Silent Spring* (New York: Houghton Mifflin Harcourt, 2002), 245–61, Chapter 15, "Nature Fights Back," for a discussion on insect resistance.

MODULE 8
PLACE IN THE AUTHOR'S WORK

KEY POINTS

- Carson was already a well-established and prize-winning author before *Silent Spring* was published.

- *Silent Spring* transformed Carson into an environmental* advocate and leader unafraid to challenge authority.

- Her ability to engage in political and environmental debate was limited, however, by her battle with breast cancer;* she died shortly after the book's publication.

Positioning

Silent Spring, by far Rachel Carson's most controversial and well-remembered major publication, was her last. She had already had a long and well-received career as an environmental writer, however.

Carson grew up near factories whose environmental impact from smokestacks was visible on a daily basis; while her deep love of nature and her writing were intrinsically linked to this upbringing, she also had a top-tier education in zoology* and environmental science* at a time when very few women achieved degrees in the sciences.

Rachel Carson lived to see the four major books she published in her lifetime (a fifth was published after her death in 1964) become best sellers and win several national awards. *Under The Sea Wind*, originally published in 1941 and republished in 1952, was a partnership with the environmental artist Howard Frech.* It is a beautiful and mystical celebration of the wonders of life under the sea at a time before high-definition underwater cameras, photographs, and films. It established Carson as a writer at ease with literary prose and an elegant use of language.

> **❝ Miss Carson … You are the lady that started this … ❞**
> United States Senator Abraham Ribicoff, in Arlene Rodda Quaratiello,
> *Rachel Carson: A Biography*

The Sea Around Us, published in 1951, "became an overnight best seller and made Rachel Carson the voice of public science in America, an internationally recognized authority on the oceans, and established her reputation as a nature writer of first rank," says the Rachel Carson Institute.[1] This book built on Carson's use of language and her wonder about the environment but also incorporated scientific fact. This was the start of her collaboration with other leading scientific experts and put her training in government research to good use. Linda Lear,* one of Carson's most famous biographers, says, "Carson does not neglect mystery and wonder but blends imagination with fact and expert knowledge."[2] *The Sea Around Us* won the National Book Award in the United States for nonfiction in 1952, and its success was critical to Carson's later impact with *Silent Spring*. Moreover, the sales of this book provided her with the financial security to resign from her government position and pursue research and writing full time.

The Edge of the Sea, published in 1955, further built on the themes of the ocean and the environment, but also incorporated a practical user guide. This trilogy of works made Carson well known as a popular, highly regarded environmental author who made scientific fact and faraway places accessible to the household reader. *Silent Spring*, then, can be seen as a natural extension of Carson's portfolio of environmental work. It combines her graceful language with the scientific data she was qualified to investigate and interpret.

Integration
Carson's background enabled her to become an environmental writer of lasting significance. Even such hardships as the financial struggles

and family obligations that prompted her to leave academia would lead to opportunities that paved the way for her role as an author. Forced to leave university, she took up a government job that exposed her to research methods, writing demands, and a network of highly skilled government scientists, all of which directly contributed to her ability to conceive and publish *Silent Spring*. Even her battle with stage IV breast cancer gave her, according to her friends, a passion, anger, and conviction that strengthened her argument in *Silent Spring*. Carson was diagnosed with breast cancer in late 1960, but kept her diagnosis hidden for fear of being criticized on the grounds that her own illness was influencing her scientific arguments.[3]

Silent Spring integrates the most successful and defining elements of Carson's previous work, brought together in the service of a new struggle. Carson's defense of the environment, together with her stance that it is our obligation to protect it for future generations, is a direct extension of her love of nature and her childhood roots on her family's farm. As a government scientist and author, Carson received letters and telegrams from citizens throughout the country alerting her to the questionable treatment of the environment they witnessed. Her love and awe for the natural world, her desire to understand it scientifically, and her yearning to protect it, all evident individually in her previous books, come together in *Silent Spring*, supported by her own strongly felt moral obligation to bring to light concerns about pesticide use and chemical toxins.

Significance
Silent Spring was undeniably the most significant piece of work Carson ever produced, and although she was already a well-known writer with an established scientific career by the time of its publication, *Silent Spring* would forever define her along with the emergence of the broader environmental movement* of the twentieth century. Carson's own reputation continues to be tied to the work's reputation,

and the same controversies that emerged immediately after its publication continue to be hotly debated today. Separating Carson from her claims in *Silent Spring* is almost impossible. Supporters continue to hail her as an environmental savior and shudder to think what the world might have been like without her; her opponents, meanwhile, still refute her claims and ascribe to her economic hardship and the death of millions killed by malaria because her book, they claim, prevented DDT from being used to eradicate mosquitos in Africa.[4]

NOTES

1   Rachel Carson Institute, Chatham University, accessed December 12, 2015, http://www.chatham.edu/centers/rachelcarson/.

2   Linda Lear, *Rachel Carson: Witness for Nature* (New York: Macmillan, 1998), 441–65.

3   Linda Lear, RachelCarson.org, accessed December 12, 2015, http://www.rachelcarson.org/SeaAroundUs.aspx.

4   William Souder, *On a Farther Shore: The Life and Legacy of Rachel Carson* (New York: Broadway Books, 2012), 332–6.

SECTION 3
IMPACT

MODULE 9
THE FIRST RESPONSES

KEY POINTS

- Responses to *Silent Spring* came from two camps: those who heeded Carson's warnings and supported her goals; and those who sought to prove her wrong and question her qualifications and motives.

- Carson anticipated the negative reactions to the book and remained engaged with the debate until she lost her life to breast cancer less than two years after publication.

- Carson's critics had widely varying arguments, many with little factual basis.

Criticism

Rachel Carson was well aware that the publication of *Silent Spring* would bring angry criticism, particularly from the chemical industry. One manufacturer of the pesticide* DDT,* the US chemical company Velsicol, threatened to sue her publisher, Houghton Mifflin. It also accused her of being a communist.*[1] Other less dramatic critiques focused on her use of scientific material. Carson was accused of "cherry-picking," or carefully selecting data to support her claims while ignoring research that weakened her argument. Of particular note are bird count data from the notably comprehensive Christmas Bird Count conducted by the environmental* advocacy group the Audubon Society.* Even though the Audubon Society was a supporter of Carson's, its annual survey showed overall bird populations increasing at the same time that she was writing about their decline.[2]

In an article of 2012, one critic states that Carson "abused, twisted, and distorted many of the studies that she cited, in a brazen act of

> **" She knew her claims would surprise 99 out of 100 people.' "**
>
> Linda Lear,* www.rachelcarson.org

scientific dishonesty."[3] He goes on to present evidence to contest three of Carson's major claims: that DDT causes cancer* in humans; that it causes bird populations to decline; and that it damages the oceans.[4] Not surprisingly, members of the agricultural chemical industry supported such claims. Parke C. Brinkley, chief executive officer of the National Agricultural Chemicals Association,* wrote, "Any harm that is caused by the use of pesticides is greatly overcompensated by the good they do."[5]

In the *Time* article "Pesticides: The Price for Progress," published on September 28, 1962, another critic accused Carson of being too "hysterical" and "feminine."[6]

In the public eye, however, Carson was largely embraced. The *New Yorker* magazine reported that 99 percent of the hundreds of letters it received in response to its publication of sections of her work were favorable, and several members of the Senate and Congress House of Representatives read excerpts into *Congressional Record*.[7]

Responses

Although Carson succumbed to breast cancer in 1964, less than two years after publication, she made a few key appearances to defend her work. She was interviewed on *CBS Reports*, a news program aired by CBS,* one of the three largest television stations in the United States. Several of Carson's biographers reported the impact her frail appearance had on the public, and in particular on her critics. "Carson's careful way of speaking dispelled any notions that she was a shrew or some kind of zealot. Carson was so sick during filming at home in suburban Maryland that in the course of the interview, she propped her head on her hands," wrote one of Carson's biographers, Eliza Griswold,* in the *New York Times* in her 2012 article celebrating the 50th anniversary of *Silent Spring*'s publication.

Carson made an appearance at a United States Senate subcommittee hearing on pesticides in 1963. President John F. Kennedy,* a supporter of Carson's, had already ordered the President's Science Advisory Committee (a body instituted to advise the president on scientific matters) to investigate the federal government's use of pesticide.[8] In the Senate hearing, Carson presented a number of policy solutions that sought to separate the regulation of chemicals from the agencies that supported and subsidized industry and agriculture. She had been hard at work on this change for many years, as she saw the intertwined interests of government and big business as part of the problem. Carson did not demand a complete ban on pesticides. She simply wanted everyone to know that these chemicals were being sprayed on their land and to be able to control their impact.[9]

Conflict and Consensus

Although popular culture and collective memory hold Carson in a special place, some critics question her legacy. Many were saddened by her death so soon after the publication of *Silent Spring*, as they felt she had much more to offer the world and many more ways to make it a cleaner, safer place. Others, though, continued to refute her claims, criticize her "doom and gloom" predictions, and question whether *Silent Spring* really had the impact its supporters claimed.

Eliza Griswold notes that the rampant use of DDT was, by the time of the book's publication, starting to reach its peak. Carson herself noted that insects* develop resistance* to specific strains of pesticides within approximately seven years, because their life and reproductive cycles are so short. So while many credit Carson for the later banning of DDT, it is unlikely that she was solely responsible, as its efficacy was already in question. Griswold also notes the beginnings of dissent against the government in early-1960s America, and while she acknowledges Carson as a forerunner in this cultural shift, she admits she was by no means alone. Other scholars directly attribute significant environmental policy victories in the United States such as the Clean Water Act,* the Clean Air Act,* and the establishment of the US government's Environmental Protection Agency* to Carson and *Silent Spring*.

The major chemical companies targeted by *Silent Spring* fought back hard, creating massive public relations campaigns outlining the benefits to human and environmental health afforded by their pesticides. These included the prevention and control of insect-borne illnesses* and the protection of food and commercial crops against infestation. Monsanto,* a large agricultural chemical manufacturer angered by Carson's claims against DDT and other chemicals it produced, published "The Desolate Year," a parody of the opening chapter of *Silent Spring*, in its corporate magazine in 1962. This evoked a fantasy world with no pesticides, with insects controlling the world and disease rampant among populations: "Imagine … the United States were to go through a single year completely without pesticides. It is under that license that we take a hard look at that desolate year, examining in some detail its devastations."[10] Carson supporters, however, refute these public relations responses, and remind us that since DDT was banned in the early 1970s, the predictions of "The Desolate Year" have never materialized.

NOTES

1   Eliza Griswold, "How 'Silent Spring' Ignited the Environmental Movement," *New York Times*, September 21, 2012.

2   Robert Zubrin, "The Truth About DDT and Silent Spring," *The New Atlantis. com*, September 27, 2012, accessed March 4, 2016, http://www. thenewatlantis.com/publications/the-truth-about-ddt-and-silent-spring.

3   Charles T. Rubin, *The Green Crusade* (Lanham, MD: Rowman & Littlefield, 1994), 38–44.

4   Zubrin, "The Truth About DDT and Silent Spring."

5   Lorus Milne and Margery Milne, "There's Poison All Around Us Now," *New York Times*, September 23, 1962.

6   Bryan Walsh, "How *Silent Spring* Became the First Shot in the War over the Environment," *Time*, September 25, 2012.

7   Milne and Milne, "There's Poison All Around Us Now."

8   "The Story of Silent Spring," Natural Resources Defense Council website, accessed December 14, 2015, http://www.nrdc.org/health/pesticides/hcarson.asp.

9   Griswold, "How 'Silent Spring.'"

10   Monsanto Corporation, "The Desolate Year," *Monsanto Magazine*, October 1962.

MODULE 10
THE EVOLVING DEBATE

KEY POINTS

- *Silent Spring* changed the way the public understood environmental* concerns.

- Environmental policymakers and debates continue to draw on Carson's work.

- Some of the biggest changes in environmental protection in the United States, including several federal laws, are credited in part to *Silent Spring*.

Uses and Problems

US President John F. Kennedy,* in office from 1961 to 1963, was a great supporter of Rachel Carson, and when *Silent Spring* was released, he instructed the President's Science Advisory Committee to research her claims about pesticides.* The committee's findings were published in 1963 in the report *The Use of Pesticides*. It largely concurred with Carson's findings and encouraged the federal government to take a more aggressive role in testing toxic substances and regulating their release into the environment.[1]

Carson had been working on a number of policy recommendations she thought could rectify some of the bureaucratic issues that contributed to pesticide use. This unpublished work on environmental policy had a significant impact on legislation and was used in a series of Congressional hearings and special studies commissioned by President Kennedy. As a result, Congress amended several pieces of legislation, including the Federal Insecticide, Fungicide, and Rodenticide Act and the Food, Drug, and Cosmetic Act (fungicides are chemicals used to control fungus such as mold; rodenticides are chemicals used to control rodents, notably mice and rats). These

> ❝ We still see the effects of unfettered human
> intervention through Carson's eyes: she popularized
> modern ecology.❞
>
> Eliza Griswold,* "How 'Silent Spring' Ignited the Environmental Movement"

changes increased the rigor of chemical toxicity reviews and better protected the public from the unknown presence and impact of chemicals in their daily lives. In 1976, the Toxic Substance Control Act* required the Environmental Protection Agency* to ensure public protection from the "unreasonable risk of injury to health or the environment."[2] Eventually, all the chemical pesticides identified in *Silent Spring* either were banned or their use was greatly restricted.

One criticism of *Silent Spring* is that Carson fails to give credit to the numerous beneficial uses of pesticides, particularly in eliminating insect-borne illnesses* such as malaria (a mosquito-borne disease which leads to fever and sometimes death) and encephalitis (a disease causing inflammation of the brain which can be borne by ticks). Pesticides can also help increase agricultural yield, which in turn promotes a more efficient use of agricultural equipment, reducing emissions from plowing, threshing, and other mechanized processes.[3]

Many of Carson's supporters make the point that she never called for an all-out ban on pesticides, as many of her critics claim. Instead, she wanted their prudent, controlled use while ensuring everyone knew about their use and their side effects. William Souder,* a Carson biographer, states that "Carson did not seek to end the use of pesticides—only their heedless overuse."

Schools of Thought

There are few neutral opinions about Rachel Carson. Supporters and critics alike are passionate and tend to hold her in either an entirely negative or an entirely positive regard. The Property and Environment

Research Center,* an organization in the United States seeking market solutions to environmental problems, says "Rachel Carson is hailed as a near saint in the environmental movement."*[4] In the field of environmental education she is regarded as a brave leader, unafraid to take on the interests of big business or the federal government while rousing the public and policymakers to action about the harmful effects of chemical pesticides.

A significant group, however, continues to speak out against her, primarily using the banning of DDT as the crux of its argument. For example, the Competitive Enterprise Institute,* a free-market advocacy group based in Washington, DC states, "Today, millions of people around the world suffer the painful and often deadly effects of malaria because one person sounded a false alarm."[5] In contrast, *Time* magazine's foreign editor has written, "Carson wasn't perfect—the quality of her book is as much in its poetry as in her ability to marshal facts—but the notion that she is somehow responsible for the continued scourge of malaria in Africa is absurd."[6]

In Current Scholarship

Silent Spring celebrated its 50th anniversary in 2012, bringing Carson much renewed attention, including through the publication of *Silent Spring at 50: The False Crises of Rachel Carson*. Edited by a team of three professors all with ties to the Cato Institute,* an American libertarian* think tank with a strong focus on environmental issues, it seeks to refute many of her original claims (libertarianism is a right-wing political position according to which a government's most important task is to guarantee the liberty of the individual).

The book makes several points about *Silent Spring*. First, it argues that Carson chose to focus exclusively on the harmful effects of DDT, while disregarding its positive benefits, notably the control of the mosquito-borne disease malaria. Second, it points out that Carson ignored bird population data from the Audubon Society* that showed that many species of birds were actually increasing in population, rather than declining. Third, it claims that her data on the cancer* epidemic ignored significant statistical factors, such as an aging population and cancer cases caused by use of tobacco.[7]

Other scientific work continues to draw upon Carson's work, however. The US technology and culture scholar Edmund Russell's* *War and Nature* describes the relationship between chemical warfare and domestic pesticide use and draws directly on Carson's work. According to the British *Observer* newspaper's science editor, "Carson's warnings are still highly relevant, both in terms of the specific threat posed by DDT and its sister chemicals and to the general ecological dangers facing humanity." He goes on to cite several ecological examples of the continued presence of chemical pesticides in various forms of wildlife.[8]

Carson alerted the public to growing concerns within the scientific community and assigned them the moral obligation to question, learn, and take action. The public response following the book's publication was strong. Thousands of private citizens wrote to their local congressional representatives or senators requesting information and demanding that action be taken. Dozens of environmental advocacy groups were established in the years following *Silent Spring*'s publication. In 1970, fueled in part by Carson's work, the United States established its Environmental Protection Agency (EPA), a federal agency whose administrator is appointed directly by the president.[9] In 1980, President Jimmy Carter posthumously awarded Rachel Carson the Presidential Medal of Freedom, the highest civilian honor in the United States, for her work in bringing environmental concerns to public awareness.[10]

NOTES

1   President's Science Advisory Committee (PSAC), *The Use of Pesticides*, May 15, 1963.

2   Toxic Substance Control Act, 15 US Code Chapter 53, 1976.

3   Property and Environment Research Center, "Silent Spring at 50: Reexamining Rachel Carson's Classic," accessed December 14, 2015, http://www.perc.org/blog/silent-spring-50–reexamining-rachel-carsons-classic.

4   Property and Environment Research Center, "Silent Spring at 50."

5   Eliza Griswold, "How 'Silent Spring' Ignited the Environmental Movement," *New York Times*, September 21, 2012.

6   Bryan Walsh, "How *Silent Spring* Became the First Shot in the War on the Environment," *Time*, September 25, 2012.

7   Roger Meiners, Pierre Desrochers, and Andrew Morriss, eds, *Silent Spring at 50: The False Crises of Rachel Carson* (Washington, DC: Cato Institute, 2012).

8   Robin McKie, "Rachel Carson and the Legacy of Silent Spring," *Guardian*, May 26, 2012.

9   Epa.gov, accessed December 5, 2015.

10 Jimmy Carter, "Presidential Medal of Freedom Remarks at the Presentation Ceremony," June 9, 1980. Online by Gerhard Peters and John T. Woolley, *The American Presidency Project*, accessed December 5, 2015, http://www.presidency.ucsb.edu/ws/?pid=45389.

MODULE 11
IMPACT AND INFLUENCE TODAY

KEY POINTS

- More than 50 years after its publication, *Silent Spring* remains a significant but controversial piece of environmental* literature.
- Carson's perspective on man's interaction with nature is still heavily debated within scientific and policy circles.
- It is still a challenge for environmentalist thinkers to discern objective truth when faced with conflicting and incomplete data.

Position

Rachel Carson's *Silent Spring* is still an important work for all those who are passionate about environmental issues and policies. The validity of Carson's specific claims regarding DDT* and other chemical pesticides and their long-term consequences continues to be heavily debated; currently Carson is generally regarded as a hero, though some people do find fault with her scientific and doom-and-gloom claims. Much of the critically hostile book *Silent Spring at 50* seeks to prove, with scientific data that compete with Carson's, that her claims were false and unsubstantiated, that she ignored any evidence that weakened her argument, and that she unnecessarily frightened the American public.[1]

While DDT has been banned in the United States since the early 1970s, the dangers of chemical pesticides and, more broadly, chemicals released into the environment are still a concern. Scientists, medical doctors, policymakers, and the general public continue to analyze the issue and form their own opinions on everything from BPA-free*

> **" Carson's book was controversial before it even was a book. "**
> William Souder,* "Rachel Carson Didn't Kill Millions of Africans"

bottles (plastic bottles made without the chemical Bisphenol A, implicated in birth defects) to how much soy is appropriate in one's diet and how many birds are killed by windmills. The underlying theme in all of these discussions is that man's impact on nature has uncertain, and sometimes dire, consequences. Furthermore, it is open to question as to which individuals and institutions have the authority and right to determine regulations and set the levels of risk and exposure. Perhaps Carson's most important legacy is introducing to the public the idea that they need to think for themselves.

Interaction

In his introduction to the 1994 edition of *Silent Spring,* the then vice president of the United States Al Gore* said, "Writing about *Silent Spring* is a humbling experience for an elected official, because Rachel Carson's book provides undeniable proof that the power of an idea can be far greater than the power of politicians." He goes on to give Carson credit for engaging him in environmental concerns. Al Gore and the Intergovernmental Panel on Climate Change (IPCC)* won the Nobel Peace Prize in 2007, in part for their mainstream blockbuster hit documentary, *An Inconvenient Truth.* * The film brought the issue of climate change into popular culture and showed how industrial actions and development have shifted the earth's climate patterns since the Industrial Revolution. *

Carson's supporters and critics continue to debate the impact *Silent Spring* had on the banning of DDT. Ironically it seems that it is her critics rather than her supporters who attribute this victory to her. Carson's critics blame her for "millions of African deaths" caused by malaria because her book, they claim, prevented DDT from being used to kill mosquito populations in Africa.[2] Her supporters, on the contrary, claim DDT was already beginning to decline in popularity by the time *Silent Spring* was published, at least in the United States, because massive spraying initiatives had led to the development of a DDT-resistant strain of mosquito. In 2006, the World Health Organization,* the division of the United Nations concerned with public health, began to revisit its DDT-spraying initiatives to combat malaria in Africa, where, Carson's biographer William Souder* points out, DDT has never been banned.

Carson's ideas are still current in the ongoing debates on climate change. Environmental scholars such as Bill McKibben* and Amory Lovins* of the United States insist we must take action to reduce atmospheric carbon and our dependence on fossil fuels. Critics of geoengineering* (the field of engineering theory that focuses on planet-scale technological solutions to mitigate the impacts of climate change and sea-level rise) cite the precautionary principle*—that it is better to limit interference with the planet as we can never be sure of the ultimate effects. Meanwhile, consumers and citizens continue to demand greater information and transparency in the products they buy and the companies they support.

The Continuing Debate

Carson's biggest opponents are the usual opponents of environmental regulation—typically those focused on private property rights, capitalist* economic markets, big business, and conservative politics (capitalism is the social and economic model dominant in the West and increasingly throughout the world, in which trade and industry are conducted for private profit). Unsurprisingly, chemical manufacturers continue to refute her claims, since *Silent Spring* posed a direct threat to their livelihood. "Carson's 'you can't be too safe' standard is seen today in the 'precautionary principle' that helps to retard the adoption of superior technology that would benefit people and the environment," said Roger Meiners, distinguished professor of economics and law at the University of Texas at Arlington, senior fellow at the Property and Environment Research Center,* and one of the three main editors of *Silent Spring at 50*.[3] He believes her simplified view of risk appears to have affected the drafting of the US federal government's Clean Air Act* and Clean Water Act* that set "impossible standards in some areas not remotely related to human health or technical feasibility."

Debates about environmental regulation, economic growth, and the precautionary principle continue to dominate every global conference on environmental matters, including the conferences on climate run under the auspices of the United Nations such as the Earth Summits in Rio of 1992 and 2012, and the United Nations Conference on Sustainable Development (Rio +20). The debate will continue because there is no clear answer and the facts and circumstances continue to change. Undeniably, Carson was and remains a significant part of this conversation.

NOTES

1   Roger Meiners, Pierre Desrochers, and Andrew Morriss, eds, *Silent Spring at 50: The False Crises of Rachel Carson* (Washington, DC: Cato Institute, 2012).

2   William Souder, *On a Farther Shore: The Life and Legacy of Rachel Carson* (New York: Broadway Books, 2012), 332–5.

3   https://www.masterresource.org/silent-spring-at-50/silent-spring-at-50/.

MODULE 12
WHERE NEXT?

KEY POINTS

- *Silent Spring* shows the importance of analyzing humanity's influence on the environment.*

- Carson's quest to alert the public to industrial practices and the loopholes in environmental regulation remains relevant today.

- *Silent Spring* acts as a call to action for anyone concerned about the long-term sustainability of the planet for future generations, regardless of whether one accepts or rejects Carson's particular claims.

Potential

Rachel Carson's *Silent Spring* is still a powerful force in bringing to light concerns over pesticide use and the environment generally. Her approach remains relevant to the political and scientific challenges presented by global climate change,* the release of gasses that trap solar energy, and energy security (a nation's assured access to energy or fuel). World leaders debate extreme ideas about geoengineering* while conservationists remind us about the precautionary principle.*

Silent Spring remains alive in these debates and scholars continue to discuss Rachel Carson's impact on fields as diverse as environmental regulation, government leadership, feminism,* environmental conservation,* and morality (ethical behavior).

Silent Spring continues to be a highly esteemed piece of nonfiction literature. It was named one of the "25 Greatest Scientific Books of All Time" by *Discover* magazine,[1] and the *Guardian*, a UK newspaper, listed it as part of its "Fifty Books to Change the World."[2] It is included

> ❝ And there the two sides sit 50 years later. On one side of the environmental debate are the perceived softhearted scientists and those who would preserve the natural order; on the other are the hard pragmatists [realists] of industry and their friends in high places, the massed might of the establishment. Substitute climate change for pesticides,* and the argument plays out the same now as it did a half-century ago. ❞
>
> William Souder, "Rachel Carson Didn't Kill Millions of Africans"

in the "100 Best Nonfiction Books of the Twentieth Century," compiled by the *National Review*, and it has a place on *Time* magazine's "All Time Greatest Nonfiction Books" listing. The number of environmental science,* study, and policy programs in higher education continues to increase, and most consider *Silent Spring* required reading. Regardless of one's particular opinion about Carson's claims, *Silent Spring* is necessary reading for anyone seeking to understand modern environmental literature, policy, and culture.

Perhaps more than any specific argument offered by Carson, her most important idea is that we, as a species, can never fully understand the impact we have on the environment, and to think otherwise is naïve. The mysteries of the oceans and the nuanced, secret wonders of the natural world Carson revered can never be fully modeled, regardless of how advanced the algorithm or computing device. Some aspects of nature will always remain unknown, and Carson encourages us to embrace this wonder with respect and humility.

Future Directions

Carson's work continues to inspire heated debate and ongoing scholarly research. The 50th anniversary of *Silent Spring's* publication prompted a deluge of attention and analysis. Carson continues to challenge critics and inspire the next generation of environmental leaders. Much of the future analysis and continuation of her work will be supported by institutions dedicated to preserving her memory and, conversely, by those that challenge it.

Several nonprofit institutions have been established in her memory. The Rachel Carson Institute at Carson's alma mater, Chatham College, "continues the legacy of Chatham's most famous alumna, Rachel Carson, Class of '29, author, scientist and credited with helping form the modern environmental movement."*[3] The Institute hosts a number of programs in her memory that are designed to help continue the values, goals, and directions of her work and support future generations of environmental scholars and leaders. Similar institutions include the Silent Spring Institute, a community of "researchers dedicated to science that serves the public interest,"[4] funded by a number of notable government institutes and private foundations. Ludwig Maximillian University in Munich hosts the Rachel Carson Center for Environment and Society, there is a national wildlife refuge in coastal Maine that bears her name, and her childhood home is maintained as a museum. In short, Carson lives on in American and international cultural memory.

In contrast, there are several activist groups that continue to critique her work and what she stood for. The website *rachelwaswrong. com* continues to claim that Carson was directly responsible for the deaths of millions of Africans from malaria, due to *Silent Spring*'s call for the regulation of DDT. It states, "This website addresses the dangers associated with anti-technology views, as embodied in Rachel Carson's *Silent Spring*. Such views pervade much of modern-day environmental literature, and have actually become part of the world's conventional wisdom."[5]

Summary

Today, Rachel Carson's *Silent Spring* forms a fundamental part of environmental literature and plays a fundamental role in the larger environmental movement. For those who embrace her ideas and goals, she will continue to be a source of inspiration and leadership, reminding us to embrace the beauty, wonder, and fragility of the natural world and to treat all living creatures with respect. For her critics, she will remain someone whose ideas must be refuted and who is a testament to the idea that society can succumb to the power of graceful words. For everyone, she continues to remind us of the need to question the status quo, to think for oneself, and to speak up and out about the issues that matter most.

Silent Spring presents the reader with a chilling warning of the dangers of a highly industrialized and growth-focused society, painting a picture of a natural world that could easily be destroyed or lost if humanity continues to act without understanding the consequences of those actions. *Silent Spring* teaches the reader to embrace wonder and science at the same time, rather than see them as opposing forces, and to understand human beings and modern society as part of, not enemies of, the natural world. It teaches the reader to ask questions and understand the answers: "If, having endured much, we have at last asserted our right to know, and if, knowing, we have concluded that we are being asked to take senseless and frightening risks, then we should no longer accept the counsel of those who tell us that we must fill our world with poisonous chemicals; we should look about and see what other course is open to us."[6]

NOTES

1   "25 Greatest Scientific Books of All Time," *Discover*, December 8, 2006.

2   "Fifty Books to Change the World," *Guardian*, January 27, 2010.

3   Rachel Carson Institute, Chatham University, accessed December 12, 2015, http://www.chatham.edu/centers/rachelcarson/.

4   Silent Spring Institute, accessed December 12, 2015, http://www.
    silentspring.org/.

5   Rachel Was Wrong, accessed December 12, 2015, www.rachelwaswrong.
    org.

6   Rachel Carson, *Silent Spring* (New York: Houghton Mifflin Harcourt, 2002),
    277–8.

GLOSSARY

GLOSSARY OF TERMS

An Inconvenient Truth: a 2006 documentary film starring the then United States vice president Al Gore that reviewed the science behind climate change and its potential effects. It was award the 2007 Academy Award for best documentary film.

Atom bomb: a highly destructive nuclear weapon, developed in the 1940s; it has been used in warfare twice, in the destruction of the Japanese cities of Hiroshima and Nagasaki in August 1945.

Audubon Society: an influential conservation advocacy group, founded in the United States in 1905.

Bioaccumulation: the process of absorbing and storing any type of chemical or compound at a rate faster than it is dissipated. It can also refer to the compounding impacts of a substance as it moves up the food chain when organisms consume contaminated prey.

Biocide: any type of chemical compound capable of killing living creatures or plants. Specific to European environmental legislation, it is defined as a "chemical substance or microorganism intended to destroy, deter, render harmless, or exert a controlling effect on any harmful organism by chemical or biological means."

Biological pest control (biological control of insects): the practice of managing landscape and agricultural pests through the use of natural biological controls such as parasites, predatory insects or pathogens, which reduces the need for chemical pesticides.

Biology: the scientific study of living organisms.

Biotechnology: the use of organisms, and biological matter more generally, in the construction of tools useful to human purposes.

BPA (Bisphenol A): a chemical compound found in plastics and epoxy resins commonly used to coat the inside of food containers such as metal cans and plastic food containers and water bottles. Some research suggests that BPA can leach into food or liquid stored in BPA-lined containers, and that consumption of BPA can have adverse effects on children and unborn fetuses.

Cancer: a disease, or group of diseases, characterized by abnormal cell growth.

Capitalism: the social and economic model dominant in the West and increasingly throughout the world, in which trade and industry are conducted for private profit

Carcinogen: a substance known to cause, or have the potential to cause, cancer.

Cato Institute: a right-wing think tank, based in Washington, DC, advocating libertarian politics.

CBS: a broadcaster in the United States; the initials stand for "Colombia Broadcast System."

Clean Air Act (1963): a United States law that seeks to improve, strengthen, and accelerate programs for the prevention and abatement of air pollution.

Clean Water Act (1972): a United States law that comprehensively addresses water quality and pollution.

Communist: someone who subscribes to the political ideology of communism, which relies on the state ownership of the means of production, the collectivization of labor, and the abolition of social class.

Competitive Enterprise Institute (CEI): a think tank founded in the United States in 1984 in Washington, DC. It advocates for libertarian economic policies, through the restriction of government regulation.

Conservation movement (in America, 1890–1920): a political and social movement with a focus on conservation, environmental protection, and a celebration of national landscape resources such as National Parks and National Forests.

Crop rotation: the agricultural method of rotating the type of crop planted in a given field or garden over a given period of time; benefits include increased soil fertility and pest control.

DDT (dichloro-diphenyl-trichloroethane): an organochlorine compound with insecticidal properties that was mainly used to control mosquito-borne malaria. Although it is effective in destroying certain living things that are harmful to animals and plants, it can also be extremely dangerous to humans and the environment.

Defoliation: stripping a tree of its leaves or otherwise causing a tree to lose the majority of its leaf cover, typically through pesticide application or insect infestation.

Earth Day: begun in 1970 and held every year on April 22, Earth Day is held to draw attention to the importance of environmental concerns and protection.

Earth stewardship: a concept promoted by the Ecological Society of America that examines both socioecological change and ecosystem resilience at all scales to enhance human well-being.

Ecofeminism: a loosely defined political movement of conservation and environmental stewardship promoted by women (particularly American women) beginning in the 1960s and 1970s, that typically combines conservation advocacy and mainstream feminist thought with earth spirituality. Key figures include Mary Daly, Susan Griffin, Ellen Willis, and Rachel Carson.

Ecology: a branch of biology that studies the way groups of living things interact with each another and with their environment.

Ecosystem: a community of living organisms in a given location.

Endocrine disruptors/disruption: a category of chemicals and substances that can interrupt the natural processes of the body's endocrine hormones.

Environment: the setting or condition in which a person, plant or animal lives or operates.

Environmental conservation: the practice of using environmental resources judiciously and ensuring continued use of environmental and natural resources for future generations while preserving the health and resilience of ecosystems.

Environmental movement: the (modern) environmental movement in the United States began in the 1960s and is a political, scientific, and social movement focused on bringing attention to environmental concerns and developing strategies to improve environmental and ecological circumstances.

Environmental Protection Agency (EPA): an agency in the executive branch of the United States government focused on environmental health, quality, and regulation, with branches at the federal, state, and local level.

Environmental science: scientific study of the environment, commonly conducted by drawing on disciplines such as biology, physics, and ecology.

Feminism: the political and cultural movement associated with the struggle for equality between the sexes.

Genetics: the science dealing with genes, the biological material allowing the transmission of characteristics from generation to generation.

Geoengineering: also known as climate engineering, this refers to the modification of the earth's climate systems via artificial means in an attempt to control atmospheric carbon and reduce the predicted impacts of climate change.

Global climate change: long-term change in patterns governing the earth's weather and temperature.

Gypsy moth (*Lymantria dispar dispar*): a moth introduced into the United States from Europe. The larvae feed on the foliage of shade and other types of trees.

Industrial Revolution: a period of intense technological and industrial growth via new manufacturing processes beginning in Britain in the 1760s with the mechanization of textile manufacturing. Several climate models indicate that this was the period in which atmospheric carbon began increasing at non-natural rates.

Insect: consisting of more than a million species, this is the most diverse group of animals, representing more than half of all known living organisms.

Insect-borne illness: a disease carried and transmitted to humans by insects.

IPCC: the Intergovernmental Panel on Climate Change, a branch of the United Nations, was formed in 1988, with its primary goal being to "stabilize greenhouse gas concentrations in the atmosphere at a level that would prevent dangerous anthropogenic [human-induced] interference with the climate system."

Land ethic: a term in Aldo Leopold's book *Sand County Almanac* that calls for a new, more respectful, and more thoughtful relationship between man and the land he inhabits.

Leukemia: a group of cancers that begin in the bone marrow.

Mammal (mammalia): a type of vertebrate characterized by having hair, three distinct middle ear bones, mammary glands to nourish their young with milk, and a neocortex (the part of the cerebral cortex involved in sight and hearing). Over 4,000 distinct mammalian species have been identified.

Monoculture: the standard agricultural practice of growing a single species over a large area. It maximizes efficiency but can make crops more vulnerable to insects or diseases.

Monsanto: a multinational agricultural chemical and biotechnology company founded in the United States in 1901.

National Agricultural Chemicals Association: an organization founded in the United States and concerned with the usage, production, and environmental effects of agricultural chemicals such as fertilizers and pesticides.

National Cancer Institute (NCI): a government agency, forming one of the 11 agencies of the US government's Department of Health and Human Services. It was founded to further research into the causes of and treatments for cancer, and for the supporters of those suffering from and affected by cancer.

National Environmental Protection Act (1970): a United States law that established the White House Council on Environmental Quality and promoted the idea that environmental impacts should be given equal consideration when making policy decisions.

National Institutes of Health (NIH): a federal agency of the US Department of Health and Human Services (USDHHS) and the primary medical research agency within the federal government.

Natural pesticides: a nonchemical or homeopathic substance used for pest control, many times in conjunction with biological pest control strategies.

Pesticide: any type of substance used to control or kill pests, in an effort either to protect agricultural crops or to maintain safety and cleanliness for human inhabitation.

Precautionary principle: an approach to risk assessment that emphasizes caution in the face of uncertainty or that prevents action without adequate knowledge of the consequences.

Property and Environment Research Center: an organization founded in the United States in 1982 seeking market solutions to environmental problems.

Radiation: the emission or transmission of energy through particle movement, typically electromagnetic waves.

Resistance (to pesticides): the phenomenon of insects becoming less impacted by a given pesticide or compound. In a given insect population, some specimens will naturally be resistant to a pesticide. The release of a pesticide will first kill off the weakest individuals, thus allowing those resistant to the compound to breed with reduced competition, resulting in a growing population of pesticide-resistant individuals.

Smithsonian Institution: the world's largest group of museums and research centers, administered by the United States federal government and established in 1846.

Soviet Union: a federation of communist states that existed between 1922 and 1991, centered primarily on Russia and its neighbors in Eastern Europe and the northern half of Asia. It was the communist pole of the Cold War, with the United States as its main "rival."

Synthetic pesticides: man-made chemicals (or mixtures of chemicals) developed to prevent or destroy certain organisms (insects, weeds, or fungi) deemed "pests."

Toxic Substance Control Act (1976): a United States law administered by the Environmental Protection Agency that, in the words of the act, "regulates commerce and protects human health and the environment by requiring testing and necessary use restrictions on certain chemical substances, and for other purposes."

US Bureau of Fisheries (now the US Fish and Wildlife Service): the main conservation agency within the Department of the Interior.

US Fish and Wildlife Service: founded as the US Bureau of Fisheries, the US Fish and Wildlife Service is the main conservation agency within the United States Department of the Interior.

World Health Organization: the division of the United Nations responsible for global public health.

World War II (1939–45): a global war between two major international alliances—the Axis (Germany, Italy, Japan) and the Allies (led by the Soviet Union, the United States, and Great Britain)—and considered the deadliest conflict in human history.

Zoology: the study of animal life.

PEOPLE MENTIONED IN THE TEXT

Erin Brockovich (b. 1960) was an American legal clerk who helped bring about one of the largest lawsuits against a utility company—Pacific Gas and Electric in 1993, for its use of hexavalent chromium. Her story was popularized by the movie *Erin Brockovich*, released in 2000 with Julia Roberts in the lead role.

Ralph Waldo Emerson (1803–82) was an American poet and author.

Howard Frech was an American environmentalist artist. He worked for the *Baltimore Sun*, where Carson met him.

Albert "Al" Gore (b. 1948) was awarded the Nobel Peace Prize in 2007 for his work in climate-change activism. He served as the 45th vice president of the United States from 1993 to 2001 under President Bill Clinton.

Eliza Griswold (b. 1973) is a widely published American journalist and poet, and the author of the *New York Times* article "How 'Silent Spring' Ignited the Environmental Movement" (2012).

Wilhelm Carl Hueper (1894–1978) was the first director of the Environmental Cancer Section of the National Cancer Institute from 1938 to 1964. He became a role model for environmental scientist and author Rachel Carson, and collaborated with her on research for *Silent Spring*.

John F. Kennedy (1917–63) was a member of the Democratic Party and 35th president of the United States. He served in office from 1961 to 1963.

Linda Lear (b. 1940) is an American writer who published *Rachel Carson: Witness for Nature* in 1997.

Aldo Leopold (1887–1948) was an author and environmentalist from the United States. He is noted for his *Sand County Almanac*, a work calling for a more respectful and thoughtful relationship between human beings and the land we inhabit.

Amory Lovins (b. 1947) is an American physicist and environmental activist who is currently chief scientist at the Rocky Mountain Institute, an environmental research and consulting organization. He featured in *Time* magazine's list of the most influential people of 2009.

Mark Hamilton Lytle is professor of history and environmental studies at Bard College in New York state. Among his works is *The Gentle Subversive* (1988), an appraisal of Rachel Carson's legacy.

Bill McKibben (b. 1960) is an American author and environmentalist who cofounded 350.org, an anti-carbon organization focused on reducing the concentration of atmospheric carbon to below 350 parts per million.

Margaret Mead (1901–78) was an American anthropologist and writer focused on environmental issues in the 1960s and 1970s.

Joni Mitchell (b. 1943) is a Canadian singer-songwriter. Beginning as a folk singer in the 1960s, she moved on to more experimental work in the 1970s, drawing on the traditions of forms such as jazz and African music.

John Muir (1838–1914) was a Scottish American naturalist whose activism was instrumental in the protection of areas of wilderness such as Yosemite Valley and the Sequoia National Park in the United States. He was an early (and notably influential) author in the tradition of environmental literature.

Paul "Pauly" Müller (1899–1965) was a Swiss chemist noted as the inventor of the pesticide DDT in 1939, for which he received the 1948 Nobel Prize in Physiology or Medicine.

Theodore "Teddy" Roosevelt (1858–1919) was a member of the Republican Party and 26th president of the United States, who served in office from 1901 to 1909.

Edmund Russell is an assistant professor in the Division of Technology, Culture, and Communication in the School of Engineering and Applied Science at the University of Virginia.

Charles Schultz (1922–2000) was an award-winning American cartoonist who was best known for the *Peanuts* characters.

William Souder is an American biographer; he is the author of *On a Farther Shore: The Life and Legacy of Rachel Carson* (2013).

Harriet Beecher Stowe (1811–96) was an American author and activist in the struggle against the institution of slavery. She is noted for her novel *Uncle Tom's Cabin* (1852), which brought the plight of African American slaves to the attention of millions around the world.

Henry David Thoreau (1817–62) was an American political theorist, author, activist, and poet, noted for coining the phrase "civil disobedience" and for his book *Walden* (1854)—an early text in the tradition of modern environmental literature.

WORKS CITED

WORKS CITED

"25 Greatest Scientific Books of All Time." *Discover*, December 8, 2006.

Belsie, Laurent. "Gypsy Moths Return to Northeast: Worst Outbreak in a Decade Descends on Northeast; Entomologists Do Not Know How to Stop It." *Christian Science Monitor*, July 2, 1990.

Brinkley, Douglas. *The Wilderness Warrior: Theodore Roosevelt and the Crusade for America*. New York: Harper Perennial, 2010.

Carson, Rachel. *Under the Sea Wind*. New York: Simon & Schuster, 1941.

The Sea Around Us. Oxford: Oxford University Press, 1951.

The Edge of the Sea. Boston: Houghton Mifflin, 1955.

Silent Spring. New York: Houghton Mifflin Harcourt, 1994, 2002.

Davis, Devra. *The Secret History of the War on Cancer*. New York: Basic Books, 2007.

"Fifty Books to Change the World." *Guardian*, January 27, 2010.

Gillam, Scott. *Rachel Carson: Pioneer of Environmentalism*. Edina, MN: ABDO, 2011.

Griswold, Eliza. "How 'Silent Spring' Ignited the Environmental Movement." *New York Times*, September 21, 2012.

Hueper, Wilhelm C. "Occupational Tumors and Allied Diseases." *Occupational Tumors and Allied Diseases* (1942).

Hynes, H. Patricia. "Unfinished Business: *Silent Spring* on the 30th Anniversary of Rachel Carson's Indictment of DDT, Pesticides Still Threaten Human Life." *Los Angeles Times*, September 10, 1992.

Keller, Julia. "Works such as *Uncle Tom's Cabin* and *Silent Spring* had such a Profound Political Effect that they Became ... Art that Changed the World." *Chicago Tribune*, June 27, 1999.

Kroll, Gary. "The 'Silent Springs' of Rachel Carson: Mass Media and the Origins of Modern Environmentalism." *Public Understanding of Science* 10, no. 4 (2001): 403–20.

Lear, Linda. *Rachel Carson: Witness for Nature*. New York: Macmillan, 1998.

RachelCarson.org. Accessed December 12, 2015. http://www.rachelcarson.org/SeaAroundUs.aspx.

Leopold, Aldo. *Sand County Almanac: And Sketches Here and There*. Oxford: Oxford University Press,1949.

Lytle, Mark Hamilton. *The Gentle Subversive: Rachel Carson, Silent Spring, and the Rise of the Environmental Movement*. New York and Oxford: Oxford University Press, 2007.

McKie, Robin. "Rachel Carson and the Legacy of Silent Spring." *Guardian*, May 26, 2012.

Meiners, Roger, Pierre Desrochers, and Andrew Morriss, eds. *Silent Spring at 50: The False Crises of Rachel Carson*. Washington, DC: Cato Institute, 2012.

Milne, Lorus, and Margery Milne. "There's Poison All Around Us Now." *New York Times*, September 23, 1962.

Monsanto Corporation. "The Desolate Year." *Monsanto Magazine* (October 1962).

Moyers, Bill. *PBS Journal*, September 21, 2007. Accessed 21 February 2016. http://www.pbs.org/moyers/journal/09212007/watch.html.

Norwood, Vera L. "The Nature of Knowing: Rachel Carson and the American Environment." *Signs* (1987): 740–60.

Oelschlaeger, Max. "Emerson, Thoreau, and the Hudson River School." *Nature Transformed*, National Humanities Center. Accessed December 30, 2015. http://nationalhumanitiescenter.org/tserve/nattrans/ntwilderness/essays/preserva.htm.

President's Science Advisory Committee (PSAC). *The Use of Pesticides*, May 15, 1963.

Property and Environment Research Center. "Silent Spring at 50: Reexamining Rachel Carson's Classic." Accessed December 14, 2015. http://www.perc.org/blog/silent-spring-50–reexamining-rachel-carsons-classic.

Quaratiello, Arlene Rodda. *Rachel Carson: A Biography.* Westport, CT: Greenwood Press, 2010.

Rothman, Joshua. "Rachel Carson's Natural Histories." *New Yorker*, September 27, 2012.

Rubin, Charles T. *The Green Crusade*. Lanham, MD: Rowman & Littlefield, 1994.

Russell, Edmund. *War and Nature: Fighting Humans and Insects with Chemicals from World War I to Silent Spring*. New York: Cambridge University Press, 2001.

Schuldenfrei, Eric. *The Films of Charles and Ray Eames: A Universal Sense of Expectation*. New York: Routledge, 2015.

Souder, William. *On a Farther Shore: The Life and Legacy of Rachel Carson*. New York: Broadway Books, 2012.

"Rachel Carson Didn't Kill Millions of Africans." *Slate*, September 4, 2012.

Stowe, Harriet Beecher. *Uncle Tom's Cabin*. Leipzig: Tauchnitz, 1852.

Walsh, Bryan. "How *Silent Spring* Became the First Shot in the War over the Environment." *Time*, September 25, 2012.

Willer, Robb. "Is the Environment a Moral Cause?" *New York Times*, February 27, 2015.

Zubrin, Robert. "The Truth About DDT and Silent Spring." *The New Atlantis.com*, September 27, 2012. Accessed March 4, 2016. http://www.thenewatlantis. com/publications/the-truth-about-ddt-and-silent-spring

THE MACAT LIBRARY
BY DISCIPLINE

AFRICANA STUDIES

Chinua Achebe's *An Image of Africa: Racism in Conrad's Heart of Darkness*
W. E. B. Du Bois's *The Souls of Black Folk*
Zora Neale Huston's *Characteristics of Negro Expression*
Martin Luther King Jr's *Why We Can't Wait*
Toni Morrison's *Playing in the Dark: Whiteness in the American Literary Imagination*

ANTHROPOLOGY

Arjun Appadurai's *Modernity at Large: Cultural Dimensions of Globalisation*
Philippe Ariès's *Centuries of Childhood*
Franz Boas's *Race, Language and Culture*
Kim Chan & Renée Mauborgne's *Blue Ocean Strategy*
Jared Diamond's *Guns, Germs & Steel: the Fate of Human Societies*
Jared Diamond's *Collapse: How Societies Choose to Fail or Survive*
E. E. Evans-Pritchard's *Witchcraft, Oracles and Magic Among the Azande*
James Ferguson's *The Anti-Politics Machine*
Clifford Geertz's *The Interpretation of Cultures*
David Graeber's *Debt: the First 5000 Years*
Karen Ho's *Liquidated: An Ethnography of Wall Street*
Geert Hofstede's *Culture's Consequences: Comparing Values, Behaviors, Institutes and Organizations across Nations*
Claude Lévi-Strauss's *Structural Anthropology*
Jay Macleod's *Ain't No Makin' It: Aspirations and Attainment in a Low-Income Neighborhood*
Saba Mahmood's *The Politics of Piety: The Islamic Revival and the Feminist Subject*
Marcel Mauss's *The Gift*

BUSINESS

Jean Lave & Etienne Wenger's *Situated Learning*
Theodore Levitt's *Marketing Myopia*
Burton G. Malkiel's *A Random Walk Down Wall Street*
Douglas McGregor's *The Human Side of Enterprise*
Michael Porter's *Competitive Strategy: Creating and Sustaining Superior Performance*
John Kotter's *Leading Change*
C. K. Prahalad & Gary Hamel's *The Core Competence of the Corporation*

CRIMINOLOGY

Michelle Alexander's *The New Jim Crow: Mass Incarceration in the Age of Colorblindness*
Michael R. Gottfredson & Travis Hirschi's *A General Theory of Crime*
Richard Herrnstein & Charles A. Murray's *The Bell Curve: Intelligence and Class Structure in American Life*
Elizabeth Loftus's *Eyewitness Testimony*
Jay Macleod's *Ain't No Makin' It: Aspirations and Attainment in a Low-Income Neighborhood*
Philip Zimbardo's *The Lucifer Effect*

ECONOMICS

Janet Abu-Lughod's *Before European Hegemony*
Ha-Joon Chang's *Kicking Away the Ladder*
David Brion Davis's *The Problem of Slavery in the Age of Revolution*
Milton Friedman's *The Role of Monetary Policy*
Milton Friedman's *Capitalism and Freedom*
David Graeber's *Debt: the First 5000 Years*
Friedrich Hayek's *The Road to Serfdom*
Karen Ho's *Liquidated: An Ethnography of Wall Street*

John Maynard Keynes's *The General Theory of Employment, Interest and Money*
Charles P. Kindleberger's *Manias, Panics and Crashes*
Robert Lucas's *Why Doesn't Capital Flow from Rich to Poor Countries?*
Burton G. Malkiel's *A Random Walk Down Wall Street*
Thomas Robert Malthus's *An Essay on the Principle of Population*
Karl Marx's *Capital*
Thomas Piketty's *Capital in the Twenty-First Century*
Amartya Sen's *Development as Freedom*
Adam Smith's *The Wealth of Nations*
Nassim Nicholas Taleb's *The Black Swan: The Impact of the Highly Improbable*
Amos Tversky's & Daniel Kahneman's *Judgment under Uncertainty: Heuristics and Biases*
Mahbub Ul Haq's *Reflections on Human Development*
Max Weber's *The Protestant Ethic and the Spirit of Capitalism*

FEMINISM AND GENDER STUDIES

Judith Butler's *Gender Trouble*
Simone De Beauvoir's *The Second Sex*
Michel Foucault's *History of Sexuality*
Betty Friedan's *The Feminine Mystique*
Saba Mahmood's *The Politics of Piety: The Islamic Revival and the Feminist Subject*
Joan Wallach Scott's *Gender and the Politics of History*
Mary Wollstonecraft's *A Vindication of the Rights of Woman*
Virginia Woolf's *A Room of One's Own*

GEOGRAPHY

The Brundtland Report's *Our Common Future*
Rachel Carson's *Silent Spring*
Charles Darwin's *On the Origin of Species*
James Ferguson's *The Anti-Politics Machine*
Jane Jacobs's *The Death and Life of Great American Cities*
James Lovelock's *Gaia: A New Look at Life on Earth*
Amartya Sen's *Development as Freedom*
Mathis Wackernagel & William Rees's *Our Ecological Footprint*

HISTORY

Janet Abu-Lughod's *Before European Hegemony*
Benedict Anderson's *Imagined Communities*
Bernard Bailyn's *The Ideological Origins of the American Revolution*
Hanna Batatu's *The Old Social Classes And The Revolutionary Movements Of Iraq*
Christopher Browning's *Ordinary Men: Reserve Police Batallion 101 and the Final Solution in Poland*
Edmund Burke's *Reflections on the Revolution in France*
William Cronon's *Nature's Metropolis: Chicago And The Great West*
Alfred W. Crosby's *The Columbian Exchange*
Hamid Dabashi's *Iran: A People Interrupted*
David Brion Davis's *The Problem of Slavery in the Age of Revolution*
Nathalie Zemon Davis's *The Return of Martin Guerre*
Jared Diamond's *Guns, Germs & Steel: the Fate of Human Societies*
Frank Dikotter's *Mao's Great Famine*
John W Dower's *War Without Mercy: Race And Power In The Pacific War*
W. E. B. Du Bois's *The Souls of Black Folk*
Richard J. Evans's *In Defence of History*
Lucien Febvre's *The Problem of Unbelief in the 16th Century*
Sheila Fitzpatrick's *Everyday Stalinism*

Eric Foner's *Reconstruction: America's Unfinished Revolution, 1863-1877*
Michel Foucault's *Discipline and Punish*
Michel Foucault's *History of Sexuality*
Francis Fukuyama's *The End of History and the Last Man*
John Lewis Gaddis's *We Now Know: Rethinking Cold War History*
Ernest Gellner's *Nations and Nationalism*
Eugene Genovese's *Roll, Jordan, Roll: The World the Slaves Made*
Carlo Ginzburg's *The Night Battles*
Daniel Goldhagen's *Hitler's Willing Executioners*
Jack Goldstone's *Revolution and Rebellion in the Early Modern World*
Antonio Gramsci's *The Prison Notebooks*
Alexander Hamilton, John Jay & James Madison's *The Federalist Papers*
Christopher Hill's *The World Turned Upside Down*
Carole Hillenbrand's *The Crusades: Islamic Perspectives*
Thomas Hobbes's *Leviathan*
Eric Hobsbawm's *The Age Of Revolution*
John A. Hobson's *Imperialism: A Study*
Albert Hourani's *History of the Arab Peoples*
Samuel P. Huntington's *The Clash of Civilizations and the Remaking of World Order*
C. L. R. James's *The Black Jacobins*
Tony Judt's *Postwar: A History of Europe Since 1945*
Ernst Kantorowicz's *The King's Two Bodies: A Study in Medieval Political Theology*
Paul Kennedy's *The Rise and Fall of the Great Powers*
Ian Kershaw's *The "Hitler Myth": Image and Reality in the Third Reich*
John Maynard Keynes's *The General Theory of Employment, Interest and Money*
Charles P. Kindleberger's *Manias, Panics and Crashes*
Martin Luther King Jr's *Why We Can't Wait*
Henry Kissinger's *World Order: Reflections on the Character of Nations and the Course of History*
Thomas Kuhn's *The Structure of Scientific Revolutions*
Georges Lefebvre's *The Coming of the French Revolution*
John Locke's *Two Treatises of Government*
Niccolò Machiavelli's *The Prince*
Thomas Robert Malthus's *An Essay on the Principle of Population*
Mahmood Mamdani's *Citizen and Subject: Contemporary Africa And The Legacy Of Late Colonialism*
Karl Marx's *Capital*
Stanley Milgram's *Obedience to Authority*
John Stuart Mill's *On Liberty*
Thomas Paine's *Common Sense*
Thomas Paine's *Rights of Man*
Geoffrey Parker's *Global Crisis: War, Climate Change and Catastrophe in the Seventeenth Century*
Jonathan Riley-Smith's *The First Crusade and the Idea of Crusading*
Jean-Jacques Rousseau's *The Social Contract*
Joan Wallach Scott's *Gender and the Politics of History*
Theda Skocpol's *States and Social Revolutions*
Adam Smith's *The Wealth of Nations*
Timothy Snyder's *Bloodlands: Europe Between Hitler and Stalin*
Sun Tzu's *The Art of War*
Keith Thomas's *Religion and the Decline of Magic*
Thucydides's *The History of the Peloponnesian War*
Frederick Jackson Turner's *The Significance of the Frontier in American History*
Odd Arne Westad's *The Global Cold War: Third World Interventions And The Making Of Our Times*

LITERATURE

Chinua Achebe's *An Image of Africa: Racism in Conrad's Heart of Darkness*
Roland Barthes's *Mythologies*
Homi K. Bhabha's *The Location of Culture*
Judith Butler's *Gender Trouble*
Simone De Beauvoir's *The Second Sex*
Ferdinand De Saussure's *Course in General Linguistics*
T. S. Eliot's *The Sacred Wood: Essays on Poetry and Criticism*
Zora Neale Huston's *Characteristics of Negro Expression*
Toni Morrison's *Playing in the Dark: Whiteness in the American Literary Imagination*
Edward Said's *Orientalism*
Gayatri Chakravorty Spivak's *Can the Subaltern Speak?*
Mary Wollstonecraft's *A Vindication of the Rights of Women*
Virginia Woolf's *A Room of One's Own*

PHILOSOPHY

Elizabeth Anscombe's *Modern Moral Philosophy*
Hannah Arendt's *The Human Condition*
Aristotle's *Metaphysics*
Aristotle's *Nicomachean Ethics*
Edmund Gettier's *Is Justified True Belief Knowledge?*
Georg Wilhelm Friedrich Hegel's *Phenomenology of Spirit*
David Hume's *Dialogues Concerning Natural Religion*
David Hume's *The Enquiry for Human Understanding*
Immanuel Kant's *Religion within the Boundaries of Mere Reason*
Immanuel Kant's *Critique of Pure Reason*
Søren Kierkegaard's *The Sickness Unto Death*
Søren Kierkegaard's *Fear and Trembling*
C. S. Lewis's *The Abolition of Man*
Alasdair MacIntyre's *After Virtue*
Marcus Aurelius's *Meditations*
Friedrich Nietzsche's *On the Genealogy of Morality*
Friedrich Nietzsche's *Beyond Good and Evil*
Plato's *Republic*
Plato's *Symposium*
Jean-Jacques Rousseau's *The Social Contract*
Gilbert Ryle's *The Concept of Mind*
Baruch Spinoza's *Ethics*
Sun Tzu's *The Art of War*
Ludwig Wittgenstein's *Philosophical Investigations*

POLITICS

Benedict Anderson's *Imagined Communities*
Aristotle's *Politics*
Bernard Bailyn's *The Ideological Origins of the American Revolution*
Edmund Burke's *Reflections on the Revolution in France*
John C. Calhoun's *A Disquisition on Government*
Ha-Joon Chang's *Kicking Away the Ladder*
Hamid Dabashi's *Iran: A People Interrupted*
Hamid Dabashi's *Theology of Discontent: The Ideological Foundation of the Islamic Revolution in Iran*
Robert Dahl's *Democracy and its Critics*
Robert Dahl's *Who Governs?*
David Brion Davis's *The Problem of Slavery in the Age of Revolution*

Alexis De Tocqueville's *Democracy in America*
James Ferguson's *The Anti-Politics Machine*
Frank Dikotter's *Mao's Great Famine*
Sheila Fitzpatrick's *Everyday Stalinism*
Eric Foner's *Reconstruction: America's Unfinished Revolution, 1863-1877*
Milton Friedman's *Capitalism and Freedom*
Francis Fukuyama's *The End of History and the Last Man*
John Lewis Gaddis's *We Now Know: Rethinking Cold War History*
Ernest Gellner's *Nations and Nationalism*
David Graeber's *Debt: the First 5000 Years*
Antonio Gramsci's *The Prison Notebooks*
Alexander Hamilton, John Jay & James Madison's *The Federalist Papers*
Friedrich Hayek's *The Road to Serfdom*
Christopher Hill's *The World Turned Upside Down*
Thomas Hobbes's *Leviathan*
John A. Hobson's *Imperialism: A Study*
Samuel P. Huntington's *The Clash of Civilizations and the Remaking of World Order*
Tony Judt's *Postwar: A History of Europe Since 1945*
David C. Kang's *China Rising: Peace, Power and Order in East Asia*
Paul Kennedy's *The Rise and Fall of Great Powers*
Robert Keohane's *After Hegemony*
Martin Luther King Jr.'s *Why We Can't Wait*
Henry Kissinger's *World Order: Reflections on the Character of Nations and the Course of History*
John Locke's *Two Treatises of Government*
Niccolò Machiavelli's *The Prince*
Thomas Robert Malthus's *An Essay on the Principle of Population*
Mahmood Mamdani's *Citizen and Subject: Contemporary Africa And The Legacy Of Late Colonialism*
Karl Marx's *Capital*
John Stuart Mill's *On Liberty*
John Stuart Mill's *Utilitarianism*
Hans Morgenthau's *Politics Among Nations*
Thomas Paine's *Common Sense*
Thomas Paine's *Rights of Man*
Thomas Piketty's *Capital in the Twenty-First Century*
Robert D. Putman's *Bowling Alone*
John Rawls's *Theory of Justice*
Jean-Jacques Rousseau's *The Social Contract*
Theda Skocpol's *States and Social Revolutions*
Adam Smith's *The Wealth of Nations*
Sun Tzu's *The Art of War*
Henry David Thoreau's *Civil Disobedience*
Thucydides's *The History of the Peloponnesian War*
Kenneth Waltz's *Theory of International Politics*
Max Weber's *Politics as a Vocation*
Odd Arne Westad's *The Global Cold War: Third World Interventions And The Making Of Our Times*

POSTCOLONIAL STUDIES

Roland Barthes's *Mythologies*
Frantz Fanon's *Black Skin, White Masks*
Homi K. Bhabha's *The Location of Culture*
Gustavo Gutiérrez's *A Theology of Liberation*
Edward Said's *Orientalism*
Gayatri Chakravorty Spivak's *Can the Subaltern Speak?*

PSYCHOLOGY

Gordon Allport's *The Nature of Prejudice*
Alan Baddeley & Graham Hitch's *Aggression: A Social Learning Analysis*
Albert Bandura's *Aggression: A Social Learning Analysis*
Leon Festinger's *A Theory of Cognitive Dissonance*
Sigmund Freud's *The Interpretation of Dreams*
Betty Friedan's *The Feminine Mystique*
Michael R. Gottfredson & Travis Hirschi's *A General Theory of Crime*
Eric Hoffer's *The True Believer: Thoughts on the Nature of Mass Movements*
William James's *Principles of Psychology*
Elizabeth Loftus's *Eyewitness Testimony*
A. H. Maslow's *A Theory of Human Motivation*
Stanley Milgram's *Obedience to Authority*
Steven Pinker's *The Better Angels of Our Nature*
Oliver Sacks's *The Man Who Mistook His Wife For a Hat*
Richard Thaler & Cass Sunstein's *Nudge: Improving Decisions About Health, Wealth and Happiness*
Amos Tversky's *Judgment under Uncertainty: Heuristics and Biases*
Philip Zimbardo's *The Lucifer Effect*

SCIENCE

Rachel Carson's *Silent Spring*
William Cronon's *Nature's Metropolis: Chicago And The Great West*
Alfred W. Crosby's *The Columbian Exchange*
Charles Darwin's *On the Origin of Species*
Richard Dawkin's *The Selfish Gene*
Thomas Kuhn's *The Structure of Scientific Revolutions*
Geoffrey Parker's *Global Crisis: War, Climate Change and Catastrophe in the Seventeenth Century*
Mathis Wackernagel & William Rees's *Our Ecological Footprint*

SOCIOLOGY

Michelle Alexander's *The New Jim Crow: Mass Incarceration in the Age of Colorblindness*
Gordon Allport's *The Nature of Prejudice*
Albert Bandura's *Aggression: A Social Learning Analysis*
Hanna Batatu's *The Old Social Classes And The Revolutionary Movements Of Iraq*
Ha-Joon Chang's *Kicking Away the Ladder*
W. E. B. Du Bois's *The Souls of Black Folk*
Émile Durkheim's *On Suicide*
Frantz Fanon's *Black Skin, White Masks*
Frantz Fanon's *The Wretched of the Earth*
Eric Foner's *Reconstruction: America's Unfinished Revolution, 1863-1877*
Eugene Genovese's *Roll, Jordan, Roll: The World the Slaves Made*
Jack Goldstone's *Revolution and Rebellion in the Early Modern World*
Antonio Gramsci's *The Prison Notebooks*
Richard Herrnstein & Charles A Murray's *The Bell Curve: Intelligence and Class Structure in American Life*
Eric Hoffer's *The True Believer: Thoughts on the Nature of Mass Movements*
Jane Jacobs's *The Death and Life of Great American Cities*
Robert Lucas's *Why Doesn't Capital Flow from Rich to Poor Countries?*
Jay Macleod's *Ain't No Makin' It: Aspirations and Attainment in a Low Income Neighborhood*
Elaine May's *Homeward Bound: American Families in the Cold War Era*
Douglas McGregor's *The Human Side of Enterprise*
C. Wright Mills's *The Sociological Imagination*

Thomas Piketty's *Capital in the Twenty-First Century*
Robert D. Putman's *Bowling Alone*
David Riesman's *The Lonely Crowd: A Study of the Changing American Character*
Edward Said's *Orientalism*
Joan Wallach Scott's *Gender and the Politics of History*
Theda Skocpol's *States and Social Revolutions*
Max Weber's *The Protestant Ethic and the Spirit of Capitalism*

THEOLOGY

Augustine's *Confessions*
Benedict's *Rule of St Benedict*
Gustavo Gutiérrez's *A Theology of Liberation*
Carole Hillenbrand's *The Crusades: Islamic Perspectives*
David Hume's *Dialogues Concerning Natural Religion*
Immanuel Kant's *Religion within the Boundaries of Mere Reason*
Ernst Kantorowicz's *The King's Two Bodies: A Study in Medieval Political Theology*
Søren Kierkegaard's *The Sickness Unto Death*
C. S. Lewis's *The Abolition of Man*
Saba Mahmood's *The Politics of Piety: The Islamic Revival and the Feminist Subject*
Baruch Spinoza's *Ethics*
Keith Thomas's *Religion and the Decline of Magic*

COMING SOON

Chris Argyris's *The Individual and the Organisation*
Seyla Benhabib's *The Rights of Others*
Walter Benjamin's *The Work Of Art in the Age of Mechanical Reproduction*
John Berger's *Ways of Seeing*
Pierre Bourdieu's *Outline of a Theory of Practice*
Mary Douglas's *Purity and Danger*
Roland Dworkin's *Taking Rights Seriously*
James G. March's *Exploration and Exploitation in Organisational Learning*
Ikujiro Nonaka's *A Dynamic Theory of Organizational Knowledge Creation*
Griselda Pollock's *Vision and Difference*
Amartya Sen's *Inequality Re-Examined*
Susan Sontag's *On Photography*
Yasser Tabbaa's *The Transformation of Islamic Art*
Ludwig von Mises's *Theory of Money and Credit*

Macat Disciplines

Access the greatest ideas and thinkers across entire disciplines, including

Postcolonial Studies

Roland Barthes's *Mythologies*
Frantz Fanon's *Black Skin, White Masks*
Homi K. Bhabha's *The Location of Culture*
Gustavo Gutiérrez's *A Theology of Liberation*
Edward Said's *Orientalism*
Gayatri Chakravorty Spivak's *Can the Subaltern Speak?*

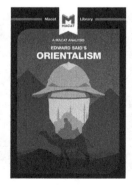

Macat analyses are available from all good bookshops and libraries.

Access hundreds of analyses through one, multimedia tool.
Join free for one month **library.macat.com**

Macat Disciplines

Access the greatest ideas and thinkers across entire disciplines, including

FEMINISM, GENDER AND QUEER STUDIES

Simone De Beauvoir's
The Second Sex

Michel Foucault's
History of Sexuality

Betty Friedan's
The Feminine Mystique

Saba Mahmood's
The Politics of Piety: The Islamic Revival and the Feminist Subject

Joan Wallach Scott's
Gender and the Politics of History

Mary Wollstonecraft's
A Vindication of the Rights of Woman

Virginia Woolf's
A Room of One's Own

Judith Butler's
Gender Trouble

Macat Disciplines

Access the greatest ideas and thinkers across entire disciplines, including

CRIMINOLOGY

Michelle Alexander's
The New Jim Crow: Mass Incarceration in the Age of Colorblindness

Michael R. Gottfredson & Travis Hirschi's
A General Theory of Crime

Elizabeth Loftus's
Eyewitness Testimony

Richard Herrnstein & Charles A. Murray's
The Bell Curve: Intelligence and Class Structure in American Life

Jay Macleod's
Ain't No Makin' It: Aspirations and Attainment in a Low-Income Neighborhood

Philip Zimbardo's
The Lucifer Effect

Macat Disciplines

Access the greatest ideas and thinkers across entire disciplines, including

GLOBALIZATION

Arjun Appadurai's, *Modernity at Large: Cultural Dimensions of Globalisation*

James Ferguson's, *The Anti-Politics Machine*

Geert Hofstede's, *Culture's Consequences*

Amartya Sen's, *Development as Freedom*

Macat Pairs

Analyse historical and modern issues from opposite sides of an argument. Pairs include:

RACE AND IDENTITY

Zora Neale Hurston's
Characteristics of Negro Expression

Using material collected on anthropological expeditions to the South, Zora Neale Hurston explains how expression in African American culture in the early twentieth century departs from the art of white America. At the time, African American art was often criticized for copying white culture. For Hurston, this criticism misunderstood how art works. European tradition views art as something fixed. But Hurston describes a creative process that is alive, ever-changing, and largely improvisational. She maintains that African American art works through a process called 'mimicry'—where an imitated object or verbal pattern, for example, is reshaped and altered until it becomes something new, novel—and worthy of attention.

Frantz Fanon's
Black Skin, White Masks

Black Skin, White Masks offers a radical analysis of the psychological effects of colonization on the colonized.

Fanon witnessed the effects of colonization first hand both in his birthplace, Martinique, and again later in life when he worked as a psychiatrist in another French colony, Algeria. His text is uncompromising in form and argument. He dissects the dehumanizing effects of colonialism, arguing that it destroys the native sense of identity, forcing people to adapt to an alien set of values— including a core belief that they are inferior. This results in deep psychological trauma.

Fanon's work played a pivotal role in the civil rights movements of the 1960s.

Macat analyses are available from all good bookshops and libraries.

Access hundreds of analyses through one, multimedia tool.
Join free for one month **library.macat.com**

Macat Pairs

Analyse historical and modern issues from opposite sides of an argument. Pairs include:

INTERNATIONAL RELATIONS IN THE 21ST CENTURY

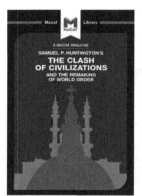

Samuel P. Huntington's
The Clash of Civilisations

In his highly influential 1996 book, Huntington offers a vision of a post-Cold War world in which conflict takes place not between competing ideologies but between cultures. The worst clash, he argues, will be between the Islamic world and the West: the West's arrogance and belief that its culture is a "gift" to the world will come into conflict with Islam's obstinacy and concern that its culture is under attack from a morally decadent "other."

Clash inspired much debate between different political schools of thought. But its greatest impact came in helping define American foreign policy in the wake of the 2001 terrorist attacks in New York and Washington.

Francis Fukuyama's
The End of History and the Last Man

Published in 1992, *The End of History and the Last Man* argues that capitalist democracy is the final destination for all societies. Fukuyama believed democracy triumphed during the Cold War because it lacks the "fundamental contradictions" inherent in communism and satisfies our yearning for freedom and equality. Democracy therefore marks the endpoint in the evolution of ideology, and so the "end of history." There will still be "events," but no fundamental change in ideology.

Macat Pairs

*Analyse historical and modern issues
from opposite sides of an argument.
Pairs include:*

HOW TO RUN AN ECONOMY

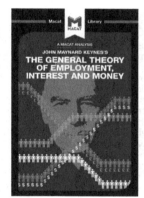

John Maynard Keynes's
*The General Theory OF Employment,
Interest and Money*

Classical economics suggests that market economies
are self-correcting in times of recession or depression,
and tend toward full employment and output. But
English economist John Maynard Keynes disagrees.

In his ground-breaking 1936 study *The General
Theory*, Keynes argues that traditional economics
has misunderstood the causes of unemployment.
Employment is not determined by the price of labor;
it is directly linked to demand. Keynes believes market
economies are by nature unstable, and so require
government intervention. Spurred on by the social
catastrophe of the Great Depression of the 1930s,
he sets out to revolutionize the way the world thinks

Milton Friedman's
The Role of Monetary Policy

Friedman's 1968 paper changed the course of
economic theory. In just 17 pages, he demolished
existing theory and outlined an effective alternate
monetary policy designed to secure 'high employment,
stable prices and rapid growth.'

Friedman demonstrated that monetary policy plays
a vital role in broader economic stability and argued
that economists got their monetary policy wrong
in the 1950s and 1960s by misunderstanding the
relationship between inflation and unemployment.
Previous generations of economists had believed
that governments could permanently decrease
unemployment by permitting inflation—and vice versa.
Friedman's most original contribution was to show that
this supposed trade-off is an illusion that only works in
the short term.

Macat analyses are available from all good bookshops and libraries.

Access hundreds of analyses through one, multimedia tool.
Join free for one month **library.macat.com**

Macat Pairs

*Analyse historical and modern issues
from opposite sides of an argument.
Pairs include:*

ARE WE FUNDAMENTALLY GOOD - OR BAD?

Steven Pinker's
The Better Angels of Our Nature
Stephen Pinker's gloriously optimistic 2011 book argues that, despite humanity's biological tendency toward violence, we are, in fact, less violent today than ever before. To prove his case, Pinker lays out pages of detailed statistical evidence. For him, much of the credit for the decline goes to the eighteenth-century Enlightenment movement, whose ideas of liberty, tolerance, and respect for the value of human life filtered down through society and affected how people thought. That psychological change led to behavioral change—and overall we became more peaceful. Critics countered that humanity could never overcome the biological urge toward violence; others argued that Pinker's statistics were flawed.

Philip Zimbardo's
The Lucifer Effect
Some psychologists believe those who commit cruelty are innately evil. Zimbardo disagrees. In *The Lucifer Effect*, he argues that sometimes good people do evil things simply because of the situations they find themselves in, citing many historical examples to illustrate his point. Zimbardo details his 1971 Stanford prison experiment, where ordinary volunteers playing guards in a mock prison rapidly became abusive. But he also describes the tortures committed by US army personnel in Iraq's Abu Ghraib prison in 2003—and how he himself testified in defence of one of those guards. committed by US army personnel in Iraq's Abu Ghraib prison in 2003—and how he himself testified in defence of one of those guards.

Macat analyses are available from all good bookshops and libraries.

Access hundreds of analyses through one, multimedia tool.
Join free for one month **library.macat.com**

Macat Pairs

Analyse historical and modern issues from opposite sides of an argument. Pairs include:

HOW WE RELATE TO EACH OTHER AND SOCIETY

Jean-Jacques Rousseau's
The Social Contract

Rousseau's famous work sets out the radical concept of the 'social contract': a give-and-take relationship between individual freedom and social order.

If people are free to do as they like, governed only by their own sense of justice, they are also vulnerable to chaos and violence. To avoid this, Rousseau proposes, they should agree to give up some freedom to benefit from the protection of social and political organization. But this deal is only just if societies are led by the collective needs and desires of the people, and able to control the private interests of individuals. For Rousseau, the only legitimate form of government is rule by the people.

Robert D. Putnam's
Bowling Alone

In *Bowling Alone*, Robert Putnam argues that Americans have become disconnected from one another and from the institutions of their common life, and investigates the consequences of this change.

Looking at a range of indicators, from membership in formal organizations to the number of invitations being extended to informal dinner parties, Putnam demonstrates that Americans are interacting less and creating less "social capital" – with potentially disastrous implications for their society.

It would be difficult to overstate the impact of *Bowling Alone*, one of the most frequently cited social science publications of the last half-century.

Printed in the United States
by Baker & Taylor Publisher Services